NEUROPHYSIOLOGY

AND EMOTION

❋

FIRST OF A SERIES ON

BIOLOGY AND BEHAVIOR

BIOLOGY AND BEHAVIOR

Neurophysiology and Emotion

Proceedings of a conference under the auspices of

Russell Sage Foundation and The Rockefeller University

David C. Glass, *Editor*

PUBLISHED BY *The Rockefeller University Press*

AND *Russell Sage Foundation* NEW YORK 1967

COPYRIGHT © 1967
BY THE ROCKEFELLER UNIVERSITY PRESS
AND RUSSELL SAGE FOUNDATION
LIBRARY OF CONGRESS
CATALOG CARD NUMBER 67-31389
PRINTED IN THE UNITED STATES OF AMERICA

Preface

THIS VOLUME contains a series of thirteen papers delivered at a conference on biology and behavior sponsored jointly by Russell Sage Foundation and The Rockefeller University. The papers have been slightly modified to meet the more formal requirements of publication. The meeting itself was held on December 10 and 11, 1965 in Caspary Auditorium at The Rockefeller University in New York City. It was designed to bring together scientists whose research combined neurophysiological and psycho-social determinants of emotional behavior.

The conference was the first in a series of three dealing with different aspects of the interrelationship between biological and behavioral systems. The second meeting was concerned primarily with behavioral genetics and was held at the Rockefeller on November 18 and 19, 1966 under the auspices of Russell Sage, the Rockefeller, and the Social Science Research Council. The third conference addressed itself to the question of environmental influences on biobehavioral systems, and was held at the Rockefeller on April 21 and 22, 1967. Volumes presenting the proceedings of each of these later conferences will appear in the near future.

The proceedings are being published primarily to serve as a general model for future exchanges between biologists and behavioral scientists. It is vitally important that both groups pursue their research with full awareness and understanding of the relevant concerns of the other. This goal may be achieved in part through conferences designed to stimulate cross-disciplinary research on basic theoretical and methodological problems that draw on biological as well as psychological and sociological concepts and methods. For this reason, the audience invited to attend the meetings were those who are actively engaged in experimentation on psychobiological or sociobiological approaches to emotional behavior. A list of conference participants may be found at the end of this volume.

Perhaps the most basic question implicit in the papers contained in this volume is the nature of the relationship between mind and body. This is a philosophical problem with strong metaphysical overtones, and has received little attention in recent years. However, much of the mind–body problem can still be seen in the present book. It is not formulated directly as an experimental hypothesis. Indeed, I suspect it cannot be framed in empirical terms. The contemporary approach is an experimental investigation of how social and psychological factors interact with physiological states to produce behavior.

The present book limits itself to emotional behavior, dealing with such topics as fear, anger, joy, euphoria, and rage. The theoretical approaches contained in this volume differ greatly, ranging from the behavioristic approach of Brady to the cognitive-physiological formulation of Schachter. Some of the papers emphasize a theory of *human* emotional behavior, others focus on the *infrahuman* level, and still others attempt a more general interspecies formulation. The volume constitutes an attempt to survey a new area of research which relies on cross-disciplinary methods and concepts. Although differing in strategy and tactics, all of the papers represent an integration of social, psychological, and biological phenomena. The ultimate goal is a better understanding of behavior through research that rises above the limitations imposed by narrow specializations.

We would like to thank Dr. Orville G. Brim, Jr., President of Russell Sage Foundation and Dr. Detlev W. Bronk, President of The Rockefeller University, whose efforts and support made the conference possible. We also want to thank Dr. Carl Pfaffmann, Vice-President of The Rockefeller University and Dr. Donald R. Young, Visiting Professor at the Rockefeller and formerly President of Russell Sage Foundation. Both men were instrumental in conceiving and implementing the idea of this conference series. The pressing need for cross-disciplinary research is reflected in Dr. Pfaffmann's behavioral sciences program at the Rockefeller and in the development of a biology-social science program at Russell Sage Foundation, including its postdoctoral fellowship program for social scientists at Rockefeller.

Russell Sage Foundation was established in 1907 by Mrs. Russell Sage for the improvement of social and living conditions in the United States. In carrying out its purpose, the Foundation conducts research under the direction of members of the staff or in close collaboration with other institutions, and supports programs designed to develop and demonstrate productive working relations between social scientists and other scientific and professional groups. The program in biology and the social sciences, which was undertaken with The Rockefeller University, represents one such activity. This book is part of the Foundation's publishing operations, which are designed to disseminate information resulting from its research and training activities. Publication under the imprint of the Foundation or, in the present case, under a joint imprint with The Rockefeller University Press, does not necessarily imply agreement by the University, the Foundation, their Trustees, or staffs with the interpretations or conclusions of the authors.

For their assistance and encouragement, I am most grateful to my colleagues at Russell Sage Foundation and The Rockefeller University. I would also like to express my gratitude to Mr. William Bayless of The Rockefeller University Press and to Mrs. Betty Davison of Russell Sage Foundation for their assistance in organizing the conferences and in publishing the proceedings. I would particularly like to thank Mrs. Helene Jordan of The Rockefeller University Press for her invaluable editorial work in bringing the present volume to publication. Finally, I would like to pay tribute to one of the conference participants, Dr. D. Wayne Woolley, who died soon after the conference. He will long be remembered by his former colleagues at Rockefeller and by the larger scientific community.

DAVID C. GLASS

Russell Sage Foundation and
The Rockefeller University

May 15, 1967

Introduction

THE TITLE of this series, "Biology and Behavior," is something of a redundancy. Over the years, behavior has been a subject of study by biological scientists—by such men as Darwin, Jennings, Pavlov, and von Bechterev, for example. Included in this early roster would be the name of Jacques Loeb, a distinguished, long-time member of the then Rockefeller Institute for Medical Research. Professor Loeb was one of the leaders of the movement to apply physico-chemical principles and methods to the study of living systems. This tradition, basic to modern biology, is fundamental to the remarkable achievements of molecular biology. Loeb approached the study of behavior from the same point of view, albeit somewhat prematurely, in his "Forced Movement, Tropism, and Animal Conduct," published in 1918. It is fitting that The Rockefeller University should return to the study of a now more sophisticated and advanced behavioral science, hopefully to initiate a period from which a unitary science of man will emerge in which his biochemistry, biophysics, and biology will be integrated with the understanding of his behavior.

Contemporary study of animal behavior has been influenced by the ethological approach, which emphasizes observation in nature for the study of instinct and motivation, although behavioral biologists have often combined experimental analysis of mechanism with field study. At the same time, psychology as a science has been influenced by the biological and physiological sciences; indeed, some psychologists regard themselves as biological scientists. It is probably in the more social wing of the psychological and social sciences that the study of behavioral phenomena and social process has been detached from the study of their biological underpinnings. The strong environmental bias of social theorists had been reinforced, in the past, by the use of overly simplified biological theories of motivation and instinct. There developed, if not an antibiological point of view, at least a nonbiological stance among students of social and cultural phenomena. Even experimental psychologists studying animal learning have reflected such attitudes. There are now signs of change in the intellectual climate.

It is in this context that the symposium series on Biology and Behavior has been organized. In each symposium we have re-examined new evidence from the biological sciences, the behavioral and social sciences, and especially from the emerging interdisciplinary fields that bridge the biological-behavioral-social domain.

It is particularly appropriate that emotion, the subject of this first volume, should be the opening vehicle of the biological-behavioral dialogue. Darwin, the father of modern evolutionary theory, wrote an extensive monograph on emotion and its expression. William James, the American psychologist, and G. C. Lange, the Dutch physiologist, propounded a theory of emotion that raised the interest of many psychologists and physiologists because it reversed common sense. Their concept that "we are afraid because we run, not that we run because we are afraid" led in part to a series of important physiological experiments on emotion by Walter B. Cannon and Philip Bard. Further studies relating endocrine and biochemical changes in emotion and their neural structures now constitute the basic foundation for the study of emotional behavior. And so the participants in the present symposium are following a well-trodden behavioral-biological path. However, the many remaining complexities are behavioral as well as neurophysiological, since environmental learning and social factors modify, modulate, and even inactivate certain underlying biological mechanisms.

This volume begins with a series of papers on a theory of emotion in the light of new observations on neural mechanisms and a review of the importance of these structures for the organization of response. Other contributors bring us up-to-date on more sophisticated observations of neuroendocrine responses to environmental and social stimuli. Counter to the strategy of searching for some fixed autonomic-endocrine response that "is the emotion" or that defines different emotional states, more recent studies show that the subtle interaction between social factors and perceptions of the emotion-producing situation are so much a part of emotion that, in fact, they change the emotion. Such studies reiterate more clearly than ever the theme that explanation of behavior cannot be made by a one-way analysis of physiological mechanism. The proper study of behavior includes both the study of physiological process and the interactions of physiology with environmental or situational determinants.

The studies in this volume on the importance of early environmental influences probably illustrate, as well as any in this area, how the subtle environmental effects often attributed to learning can, in fact, lead to relatively irreversible physiological changes. It may be these changes that provide the substrate of the environmental effect. Thus, physiological change and development set the stage for response elicited by the environment, and the response, in turn, induces physiological change that feeds back upon and influences response. Analyzing this continuing feedback between environmental factors (often mislabeled psychological and social) and biological factors (often misnamed organic) is the core problem of a biologically based behavioral science. The full appreciation of this concept is the mark of the modern study of biology and behavior.

CARL PFAFFMAN
The Rockefeller University

July 1967

Contents

Preface: DAVID C. GLASS   v

Introduction: CARL PFAFFMANN   ix

BRAIN MECHANISMS AND EMOTION

Emotion: Steps Toward a Neuropsychological Theory
     KARL H. PRIBRAM   3

The Neural Basis of Aggression in Cats
     JOHN P. FLYNN   40

Brain Mechanisms and Emotion
     RONALD MELZACK   60

ENDOCRINE SYSTEMS AND EMOTION

Emotion and Sensitivity of Psychoendocrine Systems
     JOSEPH V. BRADY   70

The Conditions for Emotional Behavior
     GEORGE MANDLER   96

Psychoendocrine Systems and Emotion: Biological Aspects
     SEYMOUR S. KETY   103

Involvement of the Hormone Serotonin in Emotion and Mind
     D. WAYNE WOOLEY   108

AUTONOMIC ACTIVITY AND EMOTION

Cognitive Effects on Bodily Functioning: Studies of Obesity and Eating
     STANLEY SCHACHTER   117

Some Psychophysiological Considerations of the Relationship Between the Autonomic Nervous System and Behavior
     MARVIN STEIN   145

Inside Every Fat Man
     NORMAN A. SCOTCH   155

CONTENTS *continued*

INFANTILE STIMULATION AND ADULT EMOTIONAL REACTIVITY

Stimulation in Infancy, Emotional Reactivity, and Exploratory Behavior
    V. H. Denenberg   161

Biology and the Emotions
    J. P. Scott   190

Analysis of Infant Stimulation
    John W. M. Whiting   201

    References   205

    Index   225

*NEUROPHYSIOLOGY
AND EMOTION*

Emotion: Steps Toward a Neuropsychological Theory

KARL H. PRIBRAM

DISCUSSION

The Neural Basis of Aggression in Cats
JOHN P. FLYNN   40

Brain Mechanism and Emotion
RONALD MELZACK   60

Do the things external which fall upon you distract you? Give yourself time to learn something new and good, and cease to be whirled around. But then you must also avoid being carried about the other way. For those too are triflers who have wearied themselves in life by their activity, and yet have no object to which to direct every movement, and, in a word, all their thoughts.

MARCUS AURELIUS: *Meditations*

INTRODUCTION

Marcus Aurelius, Rome's philosopher-emperor, developed a formula for coping with this troubled world. He pointed out that if one tries to consider problems all-of-a-piece one is overwhelmed. His prescription was simple: segment the reach of awareness; attend only to one facet of the situation at any one time; act upon that facet and then proceed to another. Too-much-too-soon is upsetting. Segmentation reduces the demands upon awareness and thereby produces imperturbability.

KARL H. PRIBRAM Stanford University School of Medicine, Palo Alto, California

This bit of wisdom can serve as the kernel for a modern neuropsychological theory of emotion. A theory crystallized from this kernel would look considerably different from those presently in vogue, yet such a theory would have to account for the popularity of these views.

Current theories of emotion—in fact, psychological theories in general—fall into two major groups: the social–behavorial, which includes the subjective, or "intrapsychic"; and the biological, which includes the physical, chemical, and, of course, the neurological. Terms are all too often taken from one frame of reference and applied to another in haphazard and uncritical fashion. In the approach presented here, every effort will be made to keep the two universes of discourse clear. The assumption is, however, that because each of these conceptual universes denotes some body of events common to both, different aspects of this common body will be illuminated by approaching it from different points of view.

Although the initial focus of the presentation is neurological, it is not made solely in neurological terms, but interdigitates the subjective-behavioral approach with the neural, and includes some aspects of the communications sciences. In this sense, the presentation is both neurobehavioral and neurocybernetic in its conception.

PREVIEW OF THE PROPOSAL

This proposal delineates the emotions as a set of processes that 1) reflect the state of relative organization or disorganization of an ordinarily stable configuration of neural systems; and 2) reflect those mechanisms which operate to redress an imbalance, not through action, but by the regulation of input. Two such mechanisms have been identified: one achieves stability by mobilizing the already available subsystems to exclude input; the other reorganizes the system to include input. These "preparatory" and "participatory" processes thus achieve control by an "internal" and an "external" route. Much of this presentation is devoted to setting forth the experimental evidence for these mechanisms and to suggest some inferences that can be drawn.

In the past, neurological theories have emphasized the relations between viscera and emotion or have linked emotion quantitatively with an amount of neural excitation. These relationships, although substantial, take into account neither the complexities of the emotional process nor the intricacies of the relevant neural apparatus. Characteristic of the latter is the hierarchy of self-regulating, equilibrating mechanisms, each of which controls its subunits but submits to regulation by a larger system. This set of systems provides the organism with stability, imperturbability.

Stated in another way, the organism's continuing stability depends on neural programs or plans — a set of genetic and experiential memory mechanisms — which organize the perceptions and behavior of the organism.[44] These programs consist of hierarchies of servomechanisms — feedback units that have been diagramed as nests of Test-Operate-Test-Exit units, or TOTE units (see Figure 1). The essential characteristics of the test mechanism is to sense incongruities, i.e., novelties; the essential characteristic of the operate mechanism is to effect changes that decrease the incongruity in the test mechanism. Input sufficiently incongruous can interrupt the ongoing plans; there is a temporary discontinuity, literally e-motion. (The word "emotion" comes from the Latin *emovere,* which means to be "out of" or "away from" motion.)

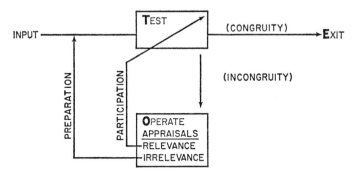

FIGURE 1 Diagram of the Test-Operate-Test-Exit (TOTE) sequence, showing the two forms (participatory and preparatory) of feedback involved in the emotional process. During participation, feedback influences the test phase to allow congruence with input; during preparation, feedback influences input to facilitate congruence with the test.

Results of experiments to be presented here suggest that the organization of emotions centers around two types of feedback which deal with the efferent control of input. They are termed *participatory* and *preparatory* processes. Participatory processes allow for the incorporation of input, and congruity is eventually achieved by alterations of the neural mechanism (or "model") against which the input is tested. On the other hand, preparatory processes alter input, thereby accomplishing congruity by manipulating the input which led to the incongruity. The system, perturbed by an incongruous input, is returned to the *status quo ante* by preparatory processes. External control is achieved by the development of new congruities, and preparation through internal control is accomplished by the return to previous congruities. Or, in terms of ongoing plans, external control develops and ramifies plans, and internal control conserves them.

This proposal differs in several respects from most currently held views on emotion. First, emphasis is on memory mechanisms, experientially derived as well as genetic, rather than on viscerally based drive processes. Second, the proposal takes as a baseline organized stability and its potential perturbation, rather than some arbitrary "level" of activation. Third, the proposal makes explicit the relation between emotion and motivation by linking both to an ongoing prebehavioral neural organization.

CURRENT NEUROLOGICAL THEORIES

There are two central themes in practically all of today's biological approaches to emotion: one deals with the relationship between visceral-glandular reactions and emotion; the other deals with the quantitative relationship between neural excitation (activation) and emotion. As already noted, these relationships do not provide an adequate framework for understanding the complexities of emotional processes nor of the intricacies of the relevant neural apparatus.

The Visceral Theme

The visceral theme is reflected everywhere in our language: "he couldn't be expected to swallow that"; "she has no stomach for it"; "he broke her heart"; "the bastard has no guts"; "he sure is bilious today." In fact, until A.D. 1800, the Galenic medical world subscribed to the notion that thoughts circulate in the ventricles of the brain, but emotions circulate in the vascular system. Gradually, medical and psychological science has become liberated from this view. But the release has been slow and guarded, partly because old theories take a long time to die and partly because the visceral view contains more than a modicum of truth. The most famous formulations that signal a retreat and liberation from the limiting aspects of the visceral concept are those of James and Lange, of Cannon and Bard, and of Papez and MacLean.

JAMES-LANGE THEORY   James and Lange faced fully the knowledge accumulated in the nineteenth century about the functions of the circulatory and nervous systems. They offered the following proposition: when an organism's reaction to a situation involves visceral structures, the sensations aroused by visceral function are perceived as emotional feelings. This proposition provoked a good deal of experimentation. A summary taken from Cannon's "Critical Examination of the James-Lange Theory of Emotions" shows the theory's weaknesses.[15]

1   Total separation of the viscera from the central nervous system does not alter emotional behavior.
2   The same visceral changes occur in very different emotional states and in nonemotional states.
3   The viscera are relatively insensitive structures.
4   Visceral changes are too slow to be a source of emotional feeling.
5   Artificial induction of the visceral changes typical of strong emotions does not produce those emotions.

CANNON-BARD THEORY   In place of the visceral theory, Cannon proposed a thalamic theory of emotions: emotional *feeling* results from stimulations of the dorsal thalamus; emotional *expression* results from the operation of hypothalamic structures. This theory was based on the observation that "sham" behavior could be elicited in decorticated and decerebrated cats except when thalamic structures also were ablated.[7] Further, a variety of expressive and visceral responses were obtained when the thalamus was electrically stimulated.[11] Finally, patients with unilateral lesions in the thalamic regions were described as sensing excessively what were to others ordinary cutaneous stimulations, e.g., a pinprick would elicit excruciating pain, warmth, intense delight.[27]

Probably more is known about the functions of these core portions of the brain than about any others. This is partly because the mechanisms are relatively "peripheral" in the sense that they can be shown to function as receptive mechanisms. Some of these structures contain receptive elements sensitive to a variety of physical and chemical agents that circulate in the blood stream and cerebrospinal fluid. In addition, the core mechanisms exert considerable direct control over the agent to which they are sensitive. This concept of control through feedback, which Cannon termed "homeostasis," has a great variety of biological and engineering applications.

Of equal importance is that the processes controlled are highly autonomous. Visceral and endocrinological regulation is performed via the sympathetic and parasympathetic portions of the autonomic nervous system. Experimental evidence was accumulated, especially by Hess,[28] to demonstrate the existence in the hypothalamic region of a trophotrophic, or energy-conserving, process that works primarily through parasympathetic peripheral divisions of the autonomic nervous system, and an ergotrophic, or mobilizing, system that works through the sympathetic division.

The balance between ergotrophic and trophotrophic is not static, of course. It could be tipped in one direction or the other; a temporary rebound, an "answering effect,"[23] could occur as the balance is restored. And indeed both processes can be activated simultaneously, so that in effect they work additively. Nor is this all. When such activation occurs, somatic as well as visceral musculature is involved.

An assumption that paralleled, if not actually guided, these studies was that an understanding of the organization of thalamically regulated processes would provide the key to an understanding of the organization of emotional processes. Once the thalamus and hypothalamus were identified as the neural substrate of emotions, this assumption followed logically.

But Lashley[37] tellingly criticized the evidence upon which this identity was assumed to rest. He pointed out that the type of disturbance on which the theory is based is as often seen to follow lesions elsewhere in the nervous system. "Hyperalgesia is not a result only of lesions within the thalamus but may arise from damage anywhere along the afferent path." He also raised the question whether "emotional disturbance" in the true sense ever occurs with thalamic lesions: "In no case was the affect referred to the source of emotional stimulation . . . but always to sensations of somatic reaction to the stimulus." He does agree that "In the hierarchy of motor centers we may recognize the thalamic region, especially the hypothalamus, as the region within which the complex patterns of expressive movements are elaborated. It does not follow from this, however, that the pathological phenomenon of hyper-excitability of emotional reactions are due solely to release from cortical inhibition or that the thalamic motor center for expressive movement contributes to the emotional experience." Clearly, the dissociation between expression and emotion, which is such a common clinical and experimental observation, can be leveled against *both* the James-Lange and the Cannon-Bard theories. Unfortunately, Lashley provided no alternative to the theories he so effectively depreciates.

PAPEZ-MACLEAN THEORY   Recently the James-Lange and Cannon-Bard views have been superseded by one proposed by Papez[48] and elaborated by MacLean.[41] The earlier theories had been firmly based on the evidence that the hypothalamus and dorsal thalamus were at the *apex* of the hierarchy of control of visceral or autonomic function. With the development of modern techniques for electrical brain stimulation, viscera were shown to be under the surveillance of the cerebral cortex. One part of this cortex—the limbic portion of the forebrain—came into focus for special attention.[29] Papez suggested that the anatomical interconnections among limbic structures were ideally constituted to handle the long-lasting, intense aspects of experience that are usually associated with emotion. To this idea MacLean added the facts of the relationship between the limbic area and viscera, and suggested that here at last is *the* visceral brain—the seat of emotions.

The persuasive power of this suggestion is great: Galen, James and Lange, Cannon and Bard, all are saved; visceral processes are the basis of emotion; an identifiable part of the brain is responsible for emotional control and experience because of its selective relations with viscera. James and Lange were wrong only in leaving out the brain; Cannon and Bard were wrong only in the part of the brain they had identified with emotion; the limbic forebrain, not the thalamus, is the responsible agent. The path from "emotions in the vascular system" to "emotions in the forebrain" had finally been completed, and each step along the way freed us from preconceptions popularly current when the step was taken.

Despite its persuasiveness and continuing popularity, there are some important criticisms of the visceral brain theory of emotions. Just as the theory gains in power from its implicit acceptance of the James-Lange and the Cannon-Bard views, so it falls heir to the criticisms leveled against the earlier theories. It has been demonstrated experimentally[75] that other parts of the cerebral mantle, when electrically stimulated, also elicit a visceral response. Emotional changes are observed to accompany lesions in parts of the forebrain other than the limbic areas. Further, ablation and stimulation of limbic structures influence problem-solving (cognitive) behavior in selective ways that cannot be attributed to changes in emotions (see below). In man, in fact, specific "memory" deficiency follows limbic lesions, while changes in "emotion" cannot be ascertained. Obviously, the Papez-MacLean theory, like its predecessors, has only a part of the problem in hand.

The Activation Theme

As one turns from the visceral to the activation theories of emotion, one can again distinguish between peripheral and central subtheories. Here, however, the argument has not been so sharp. Peripheralists have gladly accepted the diffuse, non-specific, reticular activating system as the central locus upon which and from which peripheral excitation focuses. Centralists, in turn, have been as concerned with the peripheral as with the central effects of adrenergic and cholinergic substances (e.g., Arnold[4]). Activation theory can be said, on the whole, to be less specific, less controversial, and considerably more factually oriented than visceral theories.[40] For example, a classical visceral theorist would say that a certain amount of adreno-cortical hormone circulating in the blood stream is correlated with a specific pattern of peripheral and central neural response (in hypothalamus, or visceral brain), which, in turn, corresponds to one or another emotional experience or expression. An activation theorist states merely that a correlation exists between the amount of hormone, amount of neural excitation, and amount of emotional arousal. Considerable evidence can be marshaled in favor of activation theory.

But common observation and introspection caution that something may be missing. For example, weeping is not *just* more laughing; fear is not *just* more love—although there is some truth in the notion of quantitative continuity in these processes.

A NEW APPROACH

"Expectation" and Configuration in Neural Activation

A part of the difficulty comes from the view of activation as an elementary process opposed only by another elementary process, inhibition. True, activation can be viewed as an indicator of behavioral arousal—a temporary state of disequilibrium, a perturbation of patterns of organism-environment interactions. Also, disequilibration is often sudden, explosive, and suggests agitation. But this does not necessarily mean that neural impulse transmission is facilitated: rather, a different state of organization or disorganization may suddenly have materialized. This difference is expressed as a difference in configuration, and not necessarily as a difference in the amount of neural activity. For instance, heart rate may be slowed, cortical rhythms desynchronized, peripheral blood flow diminished, but cerebral blood flow augmented. Cerebral activation, in this context, is an indicator of a configurational incongruity between input arrival patterns and established ongoing neural events.

This view of activation as an indicator of configurational change implies that the organism is fitted with a mechanism that provides a stable baseline from which such change can take off. This baseline is established by the process of habituation of the orienting reaction. Experimental evidence has accumulated in the past several years (e.g., Sokolov[68]) to show that habituation of orienting is not caused by a progressive raising of threshold to input, but to the formation of a "neuronal model" — a neuronal configuration against which subsequent inputs to the organism are matched. In essence, such neuronal configurations form the sum of an organism's expectancies. The evidence runs like this: a person is subjected to an irregular repetition of a sound stimulus of constant intensity, frequency, and duration. Initially, the person shows a set of physiological and behavioral reactions, which together form the orienting response. Among these reactions is cerebral activation, i.e., a desynchronization of the electrical rhythms recorded from the brain. As the repetition of the sound stimulus proceeds, less and less orienting takes place. For many years a simple rise in threshold to input was thought to cause such habituation. But dishabituation, i.e., a recrudescence of the orienting responses, occurs when the intensity of the sound stimulus is *de*creased, or if the duration of the sound is shortened. In this latter situation, the orienting reaction occurs as a result of the "unexpected" silence — that is, at the *off*set of the stimulation.

Each interaction between environment and organism involves at least two components: 1) discrete interaction by way of the brain's sensory-mode, specific, classical projection systems, and its core homeostats; 2) a "non-specific," relatively diffuse, interaction by way of reticular and related formations. These "non-specific" systems act as a bias on the specific reactions: the set-point or value toward which a specific interaction tends to stabilize is set by the non-specific activity. By the nature of its receptor anatomy and diffuse afferent organization, visceral feedback constitutes a major source of input to this biasing mechanism; it is an input that can do much to determine set-point. In addition, visceral and autonomic events are repetitiously redundant in the history of the organism. They vary recurrently, leading to stable habituations; this is in contrast to external changes that vary from occasion to occasion. Habituation to visceral and autonomic activity, therefore, makes up a large share, although by no means all, of the stable baseline from which the organism's reactions can take off.

Whenever the reaction to incongruous input is sufficient to disturb this baseline, the orienting reaction will include the dishabituation of visceral and autonomic activities. Such dishabituation may be subjectively felt as a mismatch between expected and actual heart rate, sweating, "butterflies," and so on. The sensing of such discrepancies is the basis for the visceral theories of emotion. The awareness of visceral-autonomic activities is thus often an indicator that a reaction to incongruous input has occurred, but it need not reflect the organization of the emotional process called forth. Support for this conception comes from the work of Lacey, et al.,[36] and of Barratt,[8,9] who have distinguished between two classes of variables in their studies of the visceral-autonomic system. One class is derived from measures of the variability of steady-state autonomic activity along a stabile-labile dimension; this dimension deals with the organism's *responsivity* — its predisposition to impulsive reaction to input. The other class of variables stems from measures of the response patterns of visceral-autonomic function.

If cerebral activation is conceived as a change in the state of organization of neural patterns related to the configurational incongruity between input and established neural activity, what then is its converse? As already indicated, over-all neuronal facilitation or inhibition are not involved. Rather, some indicator of congruity, of unperturbed, smoothly progressing neuronal activity must be sought. This indicator, at present, is found in the patterns of electrical activity recorded from the central nervous system. There is considerable evidence[1,38,39] that the slow graded activity of neural tissue, rather than the over-all inhibition or facilitation of nerve impulse transmission per se, is involved in the generation of such electrical patterns. The assumption is that the graded electrical activity recorded from brain reflects the relative stability of the neural system. Such stability would admit increments of change, provided these did not disrupt the system. Nor is it implied that incongruity and therefore activation are necessarily initiated by input. An input that ordinarily would be processed smoothly may perturb the system if that system is already unstable; or an internal change in the organism may initiate incongruity where a match had previously existed. The configuration of activation of the nervous system thus can predispose the organism toward perturbability or imperturbability.

A considerable body of evidence has recently accrued about the neurophysiological and biochemical mechanisms that regulate these predispositions. As already noted, the non-specific diffuse neural systems are primarily involved in setting the bias toward which more specific organism–environment interactions tend to stabilize. These diffuse systems are largely made up of fairly short, fine fibers with many branches. Such neuronal organizations are especially sensitive to the chemical influences in which they are immersed. A potent set of such chemical influences are the catecholamines, and they have been shown to be selectively present in the diffuse systems.[30] Further, these brain amines have been shown to be the important locus of action of the various pharmacological tranquilizers and energizers that have been so successful an adjunct in altering maladaptive emotional reaction.

Motivation and Appetite: Finickiness and Emotion

The visceral theme can also be reappraised in the light of new data. Hypothalamic mechanisms have been related both to emotion and to motivation, but the relation between these categories has never been made explicit. Hypothalamic lesions and stimulations have resulted in massive, explosive — although often directed — muscular and endocrine discharges, which are interpreted by other organisms[16,17] and by observers[15] as expressions of emotion. Hypothalamic mechanisms for the regulation of food, water, and other appetitive behavior have also been delineated in detail.[74]

However, these experiments on motivation have produced one major paradox: when lesions are made in the region of the ventromedial nucleus of the hypothalamus, rats will eat considerably more than do their controls and will become obese. But this is not all. Although rats so lesioned will eat a great deal when food is readily available, they were found to work less for food when some obstacle interfered.[45] In addition, it was found that the more palatable the food, the more the lesioned subject would eat. This gave rise to the notion that the lesioned animals were more finicky than the controls. Further, recent experimental results obtained by Krasne[35] and by Grossman[25] show that electrical stimulation of the ventromedial nucleus stops both food and water intake in the deprived rats. Yet, lesions of the ventromedial nucleus leave the rats "unmotivated" to work either for water or for food. Grossman resolves this paradox by making the suggestion that the neurobehavioral results occur due to *alterations in affect rather than appetite* when the ventromedial nucleus is manipulated: lesioned animals show an *exaggerated sensitivity to all sorts of stimulation*.

He notes that a discrepancy remains, however. Neurophysiological data show that the ventromedial and lateral hypothalamic regions form a couplet: the lateral portion serves as a feeding or "go" mechanism (which, when ablated, will produce rats that tend to starve); the ventromedial portion is a satiety, or "stop" mechanism. This remaining discrepancy is resolved if "stop" mechanisms are more generally conceived to reflect e-motion, processes ordinarily involved in taking the organism "out of motion" and thus relegating the term motivation to those processes which ordinarily result in behavior that carries forward a plan or program.

A NEUROLOGICAL MODEL AND SOME DATA

These are not the only data relevant to the problem. As already noted, autonomically regulated appetites and "tastes" are not the only concerns of a theory of emotion. Perhaps the simplest way in which to proceed is to present a neuropsychological model of the emotional process and then turn to the experiments that generated the model.

The model focuses on the interaction of two forms of afferent inhibition. The collateral type acts to *accentuate* the difference between active and less active sites; the activity of a cell inhibits the activity of its neighbors. Self-inhibition tends to *equalize* such differences; the activity of a cell inhibits its own activity through negative feedback. Any patterned change in the system will be enhanced by collateral inhibition; self inhibition works against change, tending to stabilize the status quo. Collateral inhibition is thus conceived to be a labile mechanism sensitive to input and concurrent activity. Self-inhibition, on the other hand, works more slowly, countering the rapid fluctuations in the patterns of neural activity that would otherwise occur, and stabilizing more slowly occurring changes once they have taken place. Both habituation and orienting are conceived to be dependent on the occurrence of inhibitory processes in the afferent channels of the nervous system, and both types are ubiquitous. The model states that orienting is a function of collateral inhibition and habituation a function of self-inhibition.

Neurophysiological data

The model regards inhibitory neural processes — inhibition defined as a reduction in the excitation of a neural unit — and distinguishes two major types of neural inhibition (Figure 2). Lateral, or

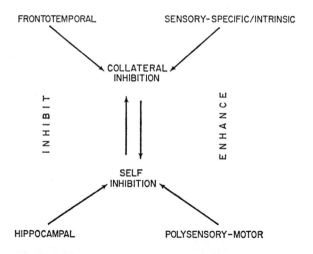

FIGURE 2 Diagram of the model of cortical control over afferent inhibitory processes.

surround, inhibition operates through collateral processes distributed among neurons or via amacrine-like cells, and is well-demonstrated in the visual,[26] auditory,[12] and somatic[47] systems, both at peripheral and central stations. This type of afferent neural interaction corresponds to Pavlov's "external" inhibition. The second type of afferent inhibition is self inhibition.[5,13,14] It takes two forms, presynaptic and postsynaptic. Both result in the regulation of afferent activity via negative feedback. In the case of postsynaptic self-inhibition, interneurons of the Renshaw type are assumed to function, via recurrent inhibitory fibers, as dampers that control the excitability of active neurons as a consequence of their own activity.

The chief concern of the model is with efferent control exerted over this interaction. This control is primarily cerebrofugal. Mechanisms that enhance and inhibit afferent inhibition are assumed to converge upon the afferent pathways. Because of this site of operation, a fourfold mechanism of efferent or cerebrofugal control should in theory be distinguishable: 1) enhancement of collateral inhibition; 2) enhancement of self-inhibition; 3) inhibition of collateral inhibition; and 4) inhibition of self-inhibition.

There is available evidence on corticofugal control over both presynaptic and postsynaptic forms of self-inhibition. Repetitive stimulation of a variety of sensory-motor points on the lateral cortex influences presynaptic inhibition at the spinal level.[2,3,22] And the effect of hippocampal stimulation on visual evoked activity has also been recorded.[24]

The evidence for efferent control of collateral inhibition has been gathered in my own laboratory, in collaboration with Dr. D. N. Spinelli.[72] Experiments were performed on fully awake monkeys implanted with small bipolar electrodes and a device that allows chronic repetitive stimulation of one of the electrode sites. The monkeys were presented with pairs of flashes and the interflash interval was varied from 25 to 200 milliseconds. Electrical responses were recorded from the striate cortex, and the amplitude of the responses was measured. A comparison of the amplitude of the second to the first response of each pair was expressed and plotted as a function. Underlying the interpretation of this function is the assumption that when the amplitude of the second pair of responses approximates that of the first, the responding cells have fully recovered their excitability. In populations of cells such as those from which these records are made, the per cent diminution of amplitude of the second response is used as an index of recovery of the total population of cells—the smaller the percentage, the fewer the number of recovered cells in the system.

Chronic stimulation (8 to 10 per second) of several cerebral sites alters this recovery function. When the inferotemporal cortex of monkeys is stimulated, recovery is delayed. Stimulation from control sites (precentral and parietal) has no such effect (Figures 3, 4, 5). Nor does the stimulation of inferotemporal cortex alter auditory recovery functions. These, however, can be changed by manipulations of the insular-temporal cortex, as was shown in a parallel experiment performed on cats. Here the crucial cortex was removed and recovery functions obtained on responses recorded from the cochlear nucleus.[19] Removal of insular-temporal cortex shortens recovery in the auditory system.

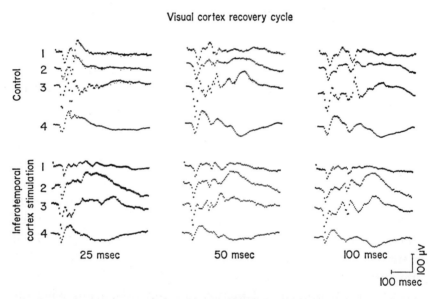

FIGURE 3 Representative record of the change produced in visual evoked responses by chronic stimulation of the inferotemporal cortex. Upper set of records was taken before stimulation; lower set, during stimulation. All traces were recorded from the visual cortex. In the first column are responses produced by a pair of flashes separated by 25 msec. Flash separation is 50 msec. in the second column and 100 msec. in the third.

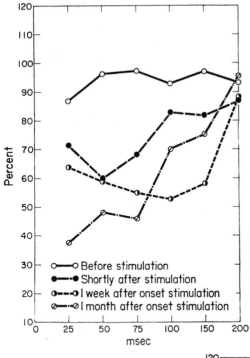

FIGURE 4 A plot of the recovery functions obtained in one monkey before and during chronic stimulation of the inferotemporal (IT) cortex.

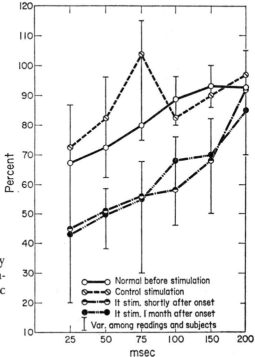

FIGURE 5 A plot of the recovery functions obtained in five monkeys before and during chronic cortical stimulation.

A great many neurobehavioral experiments have shown the importance of these isocortical temporal lobe areas (and no others) to visual and to auditory discrimination. These studies are reviewed elsewhere.[51,59] What concerns us here is that a corticofugal, efferent mechanism is demonstrated, and that this mechanism alters the rapidity with which cells in the visual and auditory afferent systems recover their excitability. Further, since stimulation delays and ablation speeds up recovery, the inference is that the normally afferent inhibitory processes which delay recovery are enhanced by the ordinary operation of these temporal lobe isocortical areas.

But the opposite effect—inhibition of afferent inhibition—can also be obtained when cerebral tissue is chronically stimulated (Figures 6, 7, 8). In these experiments the cortex of the frontal lobe and

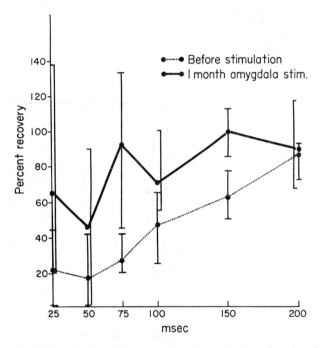

FIGURE 6 The effect on recovery function of chronic stimulation of the amygdaloid complex. The dotted line indicates the function before, the solid line after, one month of stimulation. Bars perpendicular to the curves show variability among subjects. Each curve is based on the average response of four subjects.

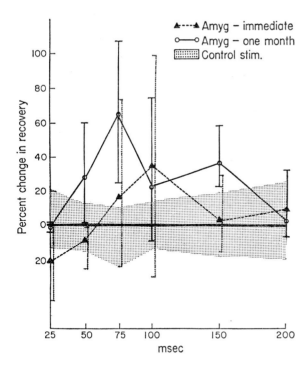

FIGURE 7 This figure represents the same data as those represented in Figure 6. However, here percentage change in recovery is plotted. Shaded area indicates range of recovery for *un*-stimulated subjects.

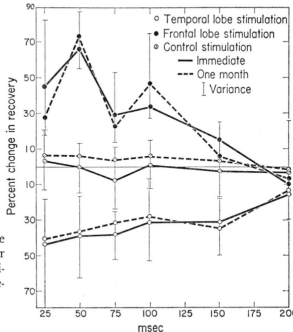

FIGURE 8 This figure plots the percentage change in recovery for all subjects in the various experiments. It is thus a summary statement of the findings.

the cortical nucleus of the amygdala were chronically stimulated, and recovery of cells in the visual system were shown to be speeded. This result has suggested that the frontal and anterior medio-basal portions of the forebrain function as efferent systems that inhibit afferent inhibitory processes.[73]

The antagonistic effect of these two efferent control systems is best illustrated by data obtained at the unit level (Figures 9, 10). Unit recordings were made from the striate cortex of flaxedilized cats to whom flashes of light were presented. Note that the silent period of a cell can be lengthened by concurrent inferotemporal stimulation. Note also that concurrent frontal stimulation can shorten this silent period. Finally, note the unit whose silent period is lengthened by inferotemporal, and shortened by frontal, stimulation.

In summary, the model is based on neurophysiological evidence of two forms of afferent inhibition: collateral and self. The reciprocal interaction of these two forms is spelled out. Data are presented indicating that afferent inhibition is under efferent corticofugal control. Further, such efferent control is shown to be balanced: both efferent enhancement and efferent inhibition of afferent inhibition were found to converge and to regulate the activity of a single system and even a single cell. The major assumption of the model is that separate forebrain systems can be found to regulate afferent neural collateral and self-inhibition.

Neurobehavioral Data

One of the consequences of this model of efferent control over afferent inhibition is a plausible neural explanation of the orienting reaction and its habituation. As already noted, a series of studies has shown: 1) that orienting can be identified by a specific pattern of behavioral and physiological indexes; and 2) that habituation of this set of indexes is not a function of a raised neural threshold to input, but to change in some neural configuration against which input is matched.[68] The reasonable suggestion can be made that habituation reflects increments of self-inhibition and that the orienting reaction manifests an override on habituation, the override taking place whenever collateral inhibition is enhanced. At least preliminary evidence at the neurophysiological level is congruent with this suggestion.[69,70,71] The following data at the neurobehavioral level can also be interpreted as in accord with the model.

The data deal with functions of the limbic forebrain.[57] Thus far, these functions have been characterized either in terms of their influence on response regulation,[43,44,53] on memory,[46,49] or on emotion.[41] In and of themselves, however, these characterizations have so far failed to provide the key to the essential nature of the limbic contribution to behavior and to psychological experience.

Bilateral amygdalectomy interferes drastically with the orienting reaction as gauged by the galvanic skin response. However, the behaviorial effect of this interference is not simple (Figures 11, 12). In a variety of discrimination learning tasks, some of which amygdalectomized monkeys found more difficult than did their controls, a behavioral measure of orienting was taken (by Dr. Patrick Bateson, an ethologist, during his post-doctoral training). This measure consisted of noting the flick of monkeys' ears during the time the cues were presented. Normal monkeys show this ear flick while they are learning; once a task has been mastered the ear response does not occur. Amygdalectomized monkeys show a greater number of such ear flicks during learning, especially in those tasks in which they have difficulty.

These results[6,10,32,34] led to the idea that orienting was made up of two components — one an alerting reaction indicated by the ear flick, the other a focusing function that allowed registration of the event producing the alerting. It is this second stage that involves the amygdala and is signaled by the appearance of a galvanic skin response.

The two phases of orienting fit the model presented. Alerting can be explained as a consequence of initial disinhibition of collateral inhibition. In the absence of a secondary controlling mechanism, this reaction would overcome the stabilizing mechanism provided by self-inhibition. Events would continually be noticed, but *adjustment* of the stabilizing mechanism (habituation) would be precluded. This is believed to be the case after amygdalectomy. By contrast, in normal subjects collateral inhibition is, in turn, inhibited by the operation of the amygdaloid mechanism. This provides the reaction with a stop mechanism that increases the likelihood that its specific configuration

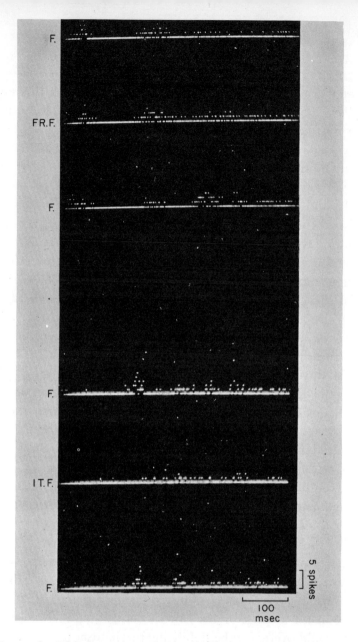

FIGURE 9 This figure is made up of photographs of a pulse histogram derived from a computer readout for average transients. Each vertical segment represents the number of impulses recorded from a neural unit during a 1.25 msec. period. The upper three traces show the effects of concurrent stimulation of the frontal cortex, the bottom three traces the effects of concurrent stimulation of the temporal cortex of cats on the unit activity evoked in the triate cortex to repeated flashes. The first and last trace in each

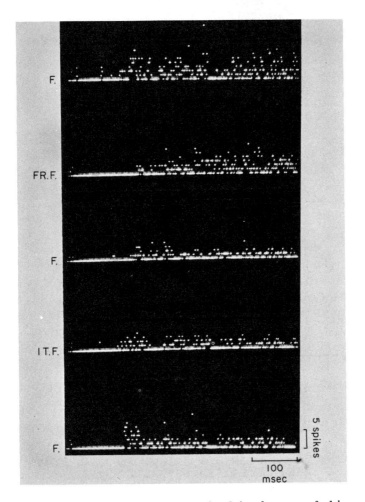

FIGURE 10 A pulse histogram obtained in the same fashion as that reproduced in Figure 9. Here the influence of concurrent frontal (second trace) and concurrent temporal (fourth trace) cortical stimulation on the flash-evoked activity of the same single unit are shown. Note that the first silent period is lengthened by concurrent frontal, and shortened by concurrent temporal, cortex stimulation.

FIGURE 9 *continued*

triad are controls; the middle traces were recorded during concurrent stimulation. Note that the first silent period is lengthened by concurrent frontal, and shortened by concurrent temporal, cortex stimulation. F, flash; FRF, frontal stimulation with flash; ITF, inferotemporal stimulation with flash.

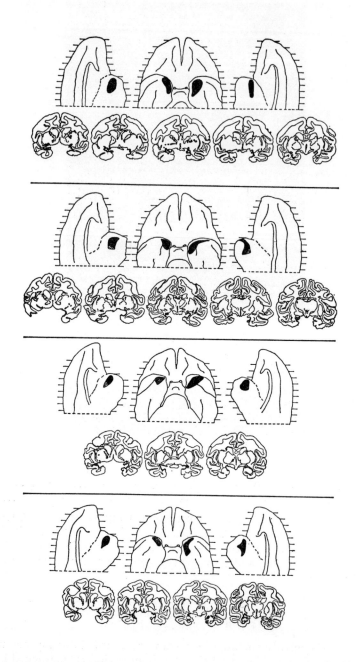

FIGURE 11 Reconstructions of the bilateral lesions of the amygdaloid complex. Black areas denote the lesion.

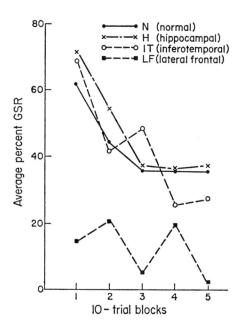

FIGURE 12 Curves of the percentage of galvanic skin response (GSR) to the first fifty presentations of the original stimulus for the normal and three experimental groups.

will be stabilized, i.e., registered. A difficulty in registering events could show up behaviorally in a variety of ways. One of them certainly would be in direct measures of habituation. Short-term measures should show an increased speed of habituation. On the other hand, longer-term measures should show that such habituation had failed to incorporate the orienting experience. This is exactly what has been found.[67] Another consequence of difficulty in registration would be the relative inefficiency of reinforcement. And, indeed, a series of experiments has shown that changing the deprivation conditions,[64] or the size of the reward,[65] or the distribution of its occurrence,[66] has considerably less effect on amygdalectomized monkeys than on their controls (Figures 13, 14, 15).

Dr. Robert Douglas, another postdoctoral student in my laboratory, first formulated in precise behavioral terms a proposal that I have taken the liberty of incorporating into my model. He suggested

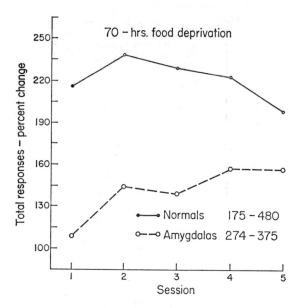

FIGURE 13 Mean percentage changes in total responses of test sessions that followed prolonged deprivation of food. The values in the legend refer to the range of total responses for the three preceding control sessions on which the percentage changes are based.

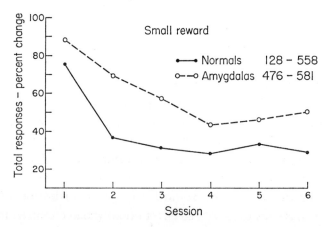

FIGURE 14 Mean percentage changes in total responses in test sessions with the small reward. (The values in this legend and in that of Figure 15 refer to the range of total responses for the three preceding large-reward sessions on which the percentage changes are based; group differences are not significant statistically.)

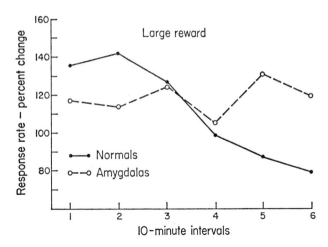

FIGURE 15 Mean percentage changes in total responses within test sessions with the large reward.

that while the amygdaloid system operates as a reinforce-register mechanism, the hippocampal formation serves as a mechanism necessary to evaluate error. He devised several ingenious experiments to test hypotheses derived from the suggestion. I shall present three of these. (For a full report, see Douglas and Pribram.[21]) All were performed in an automated discrimination apparatus, which allowed programming of tasks by a special-purpose computer that could also be used for data reduction and analysis.[58]

Dr. Douglas modified a standard behavioral testing procedure to his purpose (Figures 16, 17, 18). Ordinarily, one cue is rewarded 100 per cent of the time and the other is never rewarded. However, in the modification, called probability matching, subjects are trained to discriminate between two cues. One cue is rewarded some percentage less than 100 per cent — say 70 per cent — while the other is rewarded on the remaining occasions — in this instance, 30 per cent of the time. This task is, of course, more difficult than ordinary discriminaton, and is more interesting, because different organisms demonstrate different strategies in solving the problem. Douglas trained monkeys (bilaterally amygdalectomized, hippocampectomized, and sham-operated controls) in such a probability matching situation and then paired a

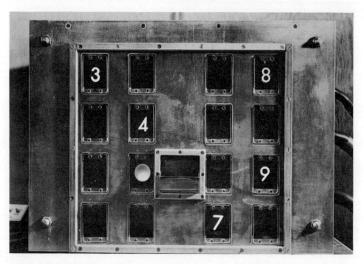

FIGURE 16 Display panel of the automated discrimination apparatus. Note 16 clear, hinged windows through which patterns can be displayed, and the central tray attached to feeder mechanism.

novel cue with either the most- or least-rewarded of the familiar cues. His results were striking.

First, monkeys with hippocampal lesions learned the probability task more slowly than did the other groups (Figure 19). This slower learning is interpreted as consonant with an impaired error-evaluate system in the hippocampectomized monkeys.

Second, monkeys with hippocampectomies, when compared with the other groups, chose the familiar cue more often when it was paired with a novel cue, irrespective of whether the familiar cue had been reinforced in 70 per cent or in 30 per cent of the trials (Figure 20). The choice of the familiar is also consonant with an intact reinforce-register function and an impaired error-evaluate mechanism.

Third, the cues used in the probability matching task were again presented, this time without reinforcement. As could be predicted, control subjects quickly shifted their responses away from the previously rewarded cues, because these responses were now erroneous. And, again, hippocampally ablated monkeys came to the support of the theory by failing to shift their responses on the basis of error (Figure 21).

FIGURE 17 Control console and special-purpose computer for the automated discrimination apparatus. This allows programing of tasks as well as data reduction and analysis.

As already noted, the behavioral process invoked to explain these results is an error-evaluate mechanism. On the basis of the model and data presented, the hippocampus is suggested as providing this mechanism. By inhibiting self-inhibition, the erroneous experience is allowed to register. In the absence of the hippocampus, the stabilizing effect of self-inhibiton is assumed to be sufficiently strong to overcome the registration of nuances: the system of afferent inhibitory processes tends to revert to the *status quo ante*. This hyperstability

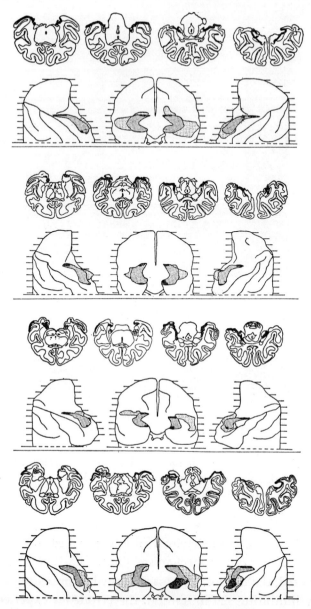

FIGURE 18 Reconstructions of the bilateral lesions of the hippocampus. Dashed areas denote the lesion, black areas denote sparing. Dotted areas show the overlying cortex removed in the approach. Heavy lines on the cross-sections show the extent of the lesion on the ventral surface.

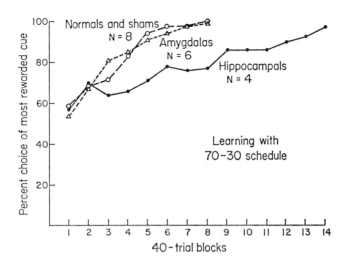

FIGURE 19 Per cent choice of most-rewarded cue in probability task involving learning with a 70 to 30 per cent reward schedule.

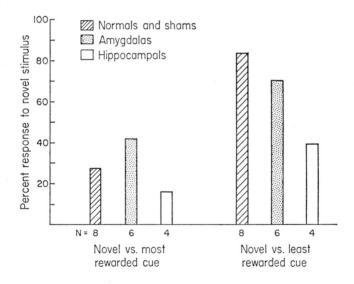

FIGURE 20 Per cent response to novel stimulus in groups of novel versus most-rewarded cue, compared with groups of novel versus least-rewarded cue.

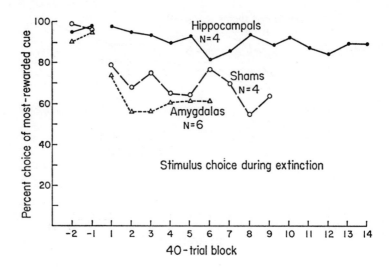

FIGURE 21 Per cent choice of most-rewarded cue in 50-trial block, where cues used in the probability matching task were again presented, this time without reinforcement.

is overcome only if the orienting events are overwhelming or if they recur regularly. Probabilistic occurrences, such as errors, fail to "get through." According to this view, short-term habituation should be slowed by hippocampectomy and registration limited to regularly recurring events. There is evidence in support of both these statements.[20,31,61]

In summary, collateral and self-afferent inhibition are bucked against one another, forming a primary couplet of neural inhibition within afferent channels. Four forebrain mechanisms are assumed to provide efferent control on this primary couplet (see Figure 2). Two of these—frontolimbic and sensory-specific-intrinsic (which includes the inferotemporal cortex)—work their influence by regulating collateral inhibition; two others—hippocampal and polysensory-motor—regulate self-inhibition. The sensory-specific-intrinsic and polysensory-motor "association" cortical systems exert their control by enhancing, while the frontotemporal and hippocampal systems exert control by inhibiting afferent neural inhibition.

THE CYBERNETICS OF EMOTION

Perturbation and the Control of Input

The significance of this demonstration of cerebral control over its own input is manifold. That this control shows two opposing tendencies is of direct relevance to the problem of emotion. One tendency accentuates orienting reactions and thus the perturbations of the system initiated by input to it. The other tendency reinforces the habituated baseline, i.e., the prior neural activity, by de-emphasizing these perturbations. In searching for adjectives for these two tendencies, the following, mentioned earlier, were deemed appropriate: *preparatory,* or better, *pre-repairatory,* and *participatory.* A preparatory process is one that prepares the organism for further interaction by repairing its perturbed system to its previous stability. A participatory process utilizes perturbation to adapt the system to the current input. Both processes are effected through feedbacks, as indicated in Figure 1. Preparatory operations are conservative and even defensive; they serve to deal with input by de-emphasis and elimination; participatory operations enhance the effect of input and so serve to increase the likelihood that the system itself will be changed.

Preparatory and participatory processes lead to different types of stability. Preparation tends to lead the organism toward relatively lasting (prospective) stability by recourse to an earlier (retrospective) organization; this type of stability is achieved through what can be termed *internal* control. Participation tolerates the temporary instabilities produced by incongruities by achieving reciprocal constancies with aspects of the environment, thus "realizing" the here-and-now (*external* control). Preparatory processes re-establish continuity at the cost of simplification. Participatory processes tolerate transience to gain flexibility through a more complex organization.

In terms of information measurement theory, these emotional processes effect a subtle balance between maximum redundancy (through preparation) and maximum information density (through participation).[62,76]

As discussed elsewhere,[53,54] the achievement of external control is conceived, through the accommodation of past experience to current input, to lead to what is subjectively felt as satisfaction. The latter occurs when similarities are identified, when congruities develop between past experience and concurrent input. On the other hand, the achievement of internal control, through the fulfillment of intentions or the restoration of ongoing plans, is conceived to result in what is experienced subjectively as gratification. The organism is gratified when there is congruity between present outcomes and past plans — when it can do things pretty much as it intended to do them. This formulation, derived from neurobehavioral data, fits the neurophysiological facts that, whereas the process labeled participation is accomplished largely through the posterior intrinsic or classical sensory "association" mechanism, the process labeled preparation is effected through the functions of the frontolimbic system.

PREPARATORY PROCESSES The clinical and experimental literature is replete with examples of preparatory processes. Concepts such as "repression," "suppression," and "perceptual defense mechanisms" can be interpreted as preparatory processes, for they are forms of defensive "gating out," that is, ignoring or repudiating those aspects of the situation that initiated the emotional state. Facets of sleep also have this "shutting out" characteristic, especially the syndromes of cataplexy and narcolepsy, which are often accompanied by affective changes.[18,34] (The above examples refer to the efferent control of afferent input at the neural level; other states represent the preparation for control of input through motivated action. Of the latter, Cannon's fight-flight reactions are probably the best known.)

By definition, preparatory processes do not accommodate the organism to the input; rather, they are internal stabilizing responses for the eradication of perturbation. The system is prepared to make itself independent of input — in effect, to make itself temporarily autonomous of concurrent input. Re-equilibration is directed toward some *status quo ante*. As it is detailed elsewhere,[60] concentration, fear, anger, and apprehension have in common the intent, implicit or explicit, to change the situation so that the organism can *repair* to the previous equilibrated state out of which it was so rudely jarred. In this way, the ongoing plans of action are conserved, provided the preparatory processes are successful in eliminating input.

But attempts to eliminate input are often not successful. The source of the disturbing input remains and the perturbation may become incessant. For when preparations are directed to the input processing channels, they have the disadvantage of not disposing of the source of the inputs responsible for disequilibration. Under such circumstances, preparations may become chronic, for the incongruities arise again and again. Repeated preparations progressively lead to the hyperstability of complete internal control; the organism becomes divorced from reality; the plans of action become inflexible. Thus, more and more, novel inputs become appraised as irrelevant, or not feasible to the ongoing plans. When this hyperstable, inflexible state is finally disrupted by an input that cannot be eliminated, the entire system becomes perturbed. Then, as the saying goes, "all hell breaks loose."

PARTICIPATORY PROCESSES   By contrast, participatory processes deal with incongruity by searching and sampling the input and accommodating the system to it. In this case, re-equilibration does not take the form of achieving the *status quo ante*; rather, the experience becomes part of the organism and the plans of action are appropriately modified. Re-equilibration, by incorporating input, proceeds to alter and restructure the organization so that it can again function gracefully, with a minimum of disequilibrium. Interest, affection, compassion, admiration, awe, wonder — all partake of this participatory quality. Such examples of participatory processes have in common some kind of involvement, engagement, or commitment to environmental events or plans that extend beyond the organism.

In the extreme, participatory reactions can lead to overwhelming external control or regulation of behavior. This makes the system highly dependent on environmental vicissitudes, with little recourse to a core organization, so the organism's ongoing plans are likely to become fragmented and the continuity of the psychological process and of behavior are sacrificed. The system becomes unstable, hyperreactive, and the organism overly distractible.

Emotion and Motivation

That an ongoing pre-perceptual and pre-behavioral organization — some dispositional context or plan — is so fundamentally related to the emotional processes of preparation and participation clarifies the relationship between motivation and emotion. Just why these two psychological categories are so often juxtaposed is seldom mentioned in the literature. When psychologists are asked to make the relationship explicit, the explanations are often muddied: "Both are related to physiological drives" (How?); "Sometimes an emotion is motivating" (In what way?); "There really isn't any difference" (Then why use both words?). However, in light of the present proposal, once it is clear that emotions are not just viscerally derived, that they stem primarily from dispositional contexts — from ongoing plans — the enigma is resolved. Motive implies action; to *e-mote* implies to be *out of* or *away from* action. In terms of the TOTE unit (Figure 1), the emotions are concerned with the regulation of input, which is to say with the feedbacks, the preparatory and participatory processes effecting efferent control over input. In essence, then, emotions are the result of neural dispositions or attitudes that regulate input when action is temporarily interrupted — literally e-motion. Motive, on the other hand, involves the organism in action, in the execution of its plans. Emotion and motivation, passion and action: these are the two poles of the Plan.

Paradoxically, whereas an organism has a good deal of control over input, it has much less control over the outcomes of behavior, except in very restricted situations. Input can be ignored, if necessary, but action always begets risk: one cannot be sure what will happen in the environment as a consequence of the action. Risk is countered only by experience.

The suggestion is that those terms we call "emotions" can also serve as names for "motives": love as an emotion has its counterpart in love as a motive. Fear the emotion has its mirror image as fear the motive. Being moved by music can be apposed to being moved to make music. And so on. Emotions and motives can, of course, be gracefully interdigitated; that is, when either the passive or active mode of the Plan becomes prepotent, maladaptation is likely to occur. Too much emotion leads either to disruption through participation or to rigidity through preparation. Furthermore, the emotion may become a disequilibrating input in itself, for it begets further incongruities which cannot be acted upon. Too much planned action, on the other hand, leads to a narrowness of purpose and a poverty in values.

In the well-constructed individual, the process of motivation and emotion go hand-in-hand. Experience is segmented, action is monitored by passion, and passion is molded into timely action. Having come full circle, this is the theory of emotion that emerges from today's neurological knowledge: the realization of Marcus Aurelius' dictum.

SUMMARY AND CONCLUSION

In conclusion, then, this proposal differs saliently in several respects from most currently held views on emotion. First, the proposal is memory-based rather than drive-based or viscerally based. Second, it takes as a baseline organized stability and its potential perturbation, rather than some level of activation. Third, it makes explicit the relation between motivation and emotion by linking both to an ongoing, prebehavioral organization, a disposition, a program, or a plan. Fourth, the proposal defines emotion as e-motion, a process which, by taking the organism out of motion, effects control not through action but by the regulation of input. Fifth, the proposal identifies, on the basis of data presented, two forms of input regulation: one reduces, the other enhances, redundancy.

Thus, two forms of such control over input are recognized. One constitutes a preparatory, protective mechanism that conserves the current configuration by simplifying the input channels and thus limiting the effective input. The other, a participatory operation, influences the input channel in the direction of complexity, thus opening the current configuration to revision by nuances.

Both preparatory and participatory processes momentarily preclude action. Control is achieved either through stabilization of the neural configuration per se or through meshing this configuration with current input. Emotion so conceived is therefore an essential mechanism for increasing the strength and flexibility of the organism's repertoire of internal alternatives with which novel situations are met.

Although different from most of the popular neurological views of emotion, this proposal does not stand completely alone. Peters[50] has made the case for emotions as appraisals—states that are related to the passive frame of reference. Melges and I have elsewhere[60] amplified and extended this relation between the proposed mechanism and subjective phenomena. On the behavioral side, Mandler has described a series of experiments in which the interruption of action is manipulated.[42] In a social context, Schachter[63] has detailed the differences between the effects of internal and external control over behavior. Although different in detail, all of these share with the present proposal the inclusion of other than visceral or activational aspects of emotion. Emotion is thus enriched, and this enrichment stems in each instance from the belated recognition by behavioral scientists that "nonaction" can be as complex and interesting a topic for psychological study as behavioral action.

The Neural Basis of Aggression in Cats

JOHN P. FLYNN

First I will criticize certain aspects of Dr. Pribram's theory of emotions, and second will present some data relevant to the neurophysiological basis of attack behavior in cats.

Dr. Pribram says that emotion means putting an organism out of action. I find this in disagreement with certain obvious forms of emotional behavior. For example, in classical "sham rage" animals, relatively trivial stimuli can release a flurry of activity—that is, a highly

---

JOHN P. FLYNN Yale University School of Medicine, New Haven, Connecticut

emotional display. Similarly, one can stimulate the hypothalamus of a cat resting quietly in a cage and cause it to attack another cat or a rat furiously. The cat is set *in* motion, not taken *out* of motion.

Dr. Pribram also says that emotions are associated with disruption of plans. To follow his usage, I think of them as sometimes *establishing* plans. Directed behavior has been a primary criterion in distinguishing "real rage" from "sham rage." If a cat chases a rat around a cage, I assume it has a plan. Certainly the rat being chased behaves as though it thought so, too.

Dr. Pribram also speaks of emotion as being associated with regulation of sensory inputs by afferent inhibition. I agree that sensory inputs are regulated in emotional behavior, but the emergence of a sensory field during stimulation of the brain is not inhibition, or even inhibition of inhibition. Later I will discuss an example of a change in the sensory field that occurs in cats stimulated to attack.

Furthermore, the changes that take place in emotional states are not restricted to variation of sensory input. Preparation for fight or flight takes place via the sympathetic nervous system, a motor system. Our own data also suggest that there are analogous preparations at the level of the nerve cells controlling skeletal musculature.

In view of this volume's intention to deal with the neurophysiological basis of emotions, I would like to state what I mean by emotion, and how evidence can be gained about its neurophysiological basis. Then I will summarize our findings with respect to the cat's attack upon a rat, and point out some of the various neurophysiological mechanisms involved in this form of behavior.

The term emotion has at least three meanings. First, it is regarded as a purely subjective feeling, recognized only through our own introspections or, secondarily, through those of other human beings — if they choose to tell us about them. The basic mechanisms within the central nervous system that mediate emotion in this sense can be investigated only in conscious human beings, and there is a relative paucity of such information. Second, emotion may mean an expression or display, which may or may not be accompanied by a subjective feeling, such as the sham rage of a decorticate dog, or the facial expression of an actor. The parts of the neural axis needed for one such display, namely sham rage, were determined by Bard[2] in 1928. Finally, certain complex forms of the organism's interaction with the environment are commonly regarded as emotional. Fight, flight, and sexual behavior fall into this category. Both the second and the third types of emotion can be studied in animals, and most of our knowledge of the physiological basis of emotions is derived from work with them.

Instead of covering the entire gamut of emotions, the present discussion is restricted to rage and attack and its neural basis. Sham rage was first described by Goltz,[8] who found that decorticate dogs were easily aroused to displays of anger by trivial stimuli, and that they were relatively unable to direct their attack. Subsequent investigations[26,27] verified Goltz' observations. Similar phenomena were extensively observed in decorticate cats by Bard and Rioch.[4]

The display presented by such animals was very much like that of a normal animal. The deficiencies lay in the animal's response to the environment. They were inadequate in directing their attacks against the offending objects. This, as well as the highly justifiable tendency on the part of the experimenter to interpret his data conservatively, led to the notion that the display was "sham" rather than "real" rage.

A display of rage can be elicited in otherwise normal cats by stimulating the hypothalamus. This phenomenon was first reported by Hess[9] in 1927 and later, but apparently independently, by Ranson[25] in 1934. J. Masserman,[21] in a series of studies starting in 1936, observed the effects of stimulating the hypothalamus in cats, and concluded that stimulating the hypothalamus was the same as exciting an efferent nerve connected to a muscle; that it was solely a motor response. His conclusion was founded on observations that the ostensibly aggressive activity was not directed toward specific objects in the environment: the cats would dash into the side of the cage, neglecting a readily available avenue of escape; all the so-called sham reactions ceased abruptly at the end of the stimulus; the animal often continued to lap milk, purr, clean its fur, or respond to petting during hypothalamic stimulation, despite a concurrent display of rage. In addition, Masserman contended that subjective experience in the cat was indicated by the formation of a conditioned response, and because eighteen out of twenty-four cats failed to show evidence of conditioning of any components of sham rage when hypothalamic stimulation was paired with light, sound, or an air puff, he concluded that there was no affective experience associated with hypothalamic stimulation.

The observations of Hess were not in accord with those of Masserman. Hess and Akert[10] stated that "a slight move on the part of the observer is sufficient to make him the object of a brisk and well-directed assault. The sharp teeth and claws are effectively utilized in this attack." Nakao[23] also saw this phenomenon, and in addition found that a stimulated cat would bite a stick placed in front of it. Hunsperger[12] made similar observations. Thus, these observers found some degree of directedness in the animal's behavior, where Masserman had seen none.

Also contrary to Masserman's general observations, Delgado, et al.,[5] showed that it was possible to condition an escape response originally elicited by stimulation of the hypothalamus. Nakao secured evidence of conditioning in his aggressive cats, as well as in the animals that escaped. Masserman himself had observed signs of conditioning in six of his cats, whose tendencies toward rage or escape were not described.

For my part, I believe that the study of directly elicitable behavior should take precedence over the study of learned responses. For this reason Wasman and I[32] decided to investigate attack itself in a systematic, rather than an incidental, fashion. We thought that a rat should be made available to a stimulated cat in a regular fashion. Control over a cat's attack upon a rat can be secured by using only those cats that do not attack rats spontaneously. In our experience, the majority of laboratory cats fall into this category. An unstimulated cat remains quietly in the cage with a rat. When the current is turned on, the cat approaches and strikes or bites the rat, stopping as soon as the current is turned off. The speed and ferocity of the attack can be controlled by the experimenter. The principal measure used to assess the cat's performance is the time from the onset of stimulation until the cat bites or strikes the rat. The response is comparable to a reflex in its regularity, although it is by no means as stereotyped as most reflexes.

The directed nature of the attack is clear, because a stimulated cat will chase a rat around the cage, and jump at it if the rat is clinging to the wall. However, an occasional cat simply shows a display of the sort described by Masserman, and does not attack either the experimenter or the rat.

When attack is the criterion, two forms can be seen. One is accompanied by a typical rage display. In the other the display is minimal. Without an object to be attacked, it would not be detected. In quiet biting attack, the cat, on being stimulated, moves swiftly about the cage with its nose low to the ground, back somewhat arched, and hair slightly on end. It usually goes directly to the rat and bites it viciously. Illustrations of these two forms are given in Figure 1.

We[1] have obtained data indicating that there is an aversive component associated with the affective attack or attack with display that is not associated with quiet biting attack. Prior to any training, we first determined as a control the number of trials out of a total of twenty on which a cat stimulated in the hypothalamus alone would go to a stool at some distance from a starting box. The cat was then trained to stop a shock to its tail by climbing on the stool. After this had been learned, the hypothalamus alone was stimulated on twenty trials, interspersed with those on which the tail was shocked. It was found that the cat would go to the stool if it were stimulated through electrodes that elicited affective display and several forms of vocalization, but not if quiet biting attack was elicited. The dichotomy is not absolute, as some cats that hissed or meowed did not go to the stool any more often than did those in which the attack was accompanied by little or no display. The data are shown in Figure 2.

The quiet biting form of attack is suggestive of hunting, because the stimulated cats often assume a stalking position. However, there is reason to hesitate in drawing this conclusion. Autonomic changes, e.g., pupillary dilation and some piloerection, do occur in conjunction with quiet biting attack, even though they are much less obvious than those in the cats showing a marked display. Furthermore, killing does not change to eating when the stimulation is continued, as one

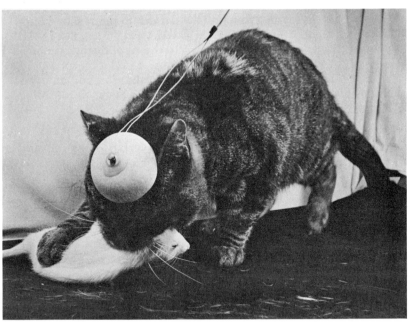

FIGURE 1 Attack, accompanied by a display of rage, is shown at the top. Quiet biting attack appears at the bottom.

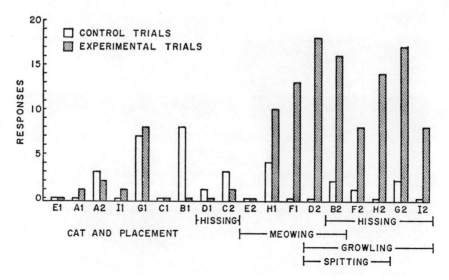

FIGURE 2 The number of trials in which a cat jumped on a stool when stimulated in the hypothalamus is given for each cat (indicated by a letter) and for each electrode (indicated by a number) before learning (control trials) and after learning (experimental trials). The cats learned to terminate a shock to the tail by jumping on the stool. After learning, the cats showed an increase in the number of trials in which they went to the stool during stimulation of the hypothalamus alone, when stimulation produced attack accompanied by several forms of vocalization and an affective display. There was no significant increase when sites leading to quiet biting attack were stimulated.

might expect in hunting. There is some overlap of eating and attack,[13] but the two are by no means inseparable. An additional reason for distinguishing between eating and attack is provided by the observation that cats starved for as long as 72 hours, and then given their first food in a cage with a rat, will break off eating to attack the rat when stimulated.

Are cats selective or indiscriminate in their choice of objects when stimulated to attack? The answer[15] to this is that they do not attack styrofoam blocks, or foam rubber blocks, but will attack anesthetized or stuffed rats. Three objects were put in the front of the cat's cage, and the cat was placed in the center toward the back. An anesthetized rat was used on every trial and so was a stuffed rat. The third object was either a styrofoam block, a foam rubber block, or a small stuffed toy dog. The object attacked on each trial was determined (Figure 3). Seven of the nine cats in the experiment never attacked the styrofoam or foam rubber blocks. Three of the animals (those on the top row in Figure 3) had a significant preference for the anesthetized rat. Five cats showed no reliable preference in attacking stuffed or anesthetized rats, while one cat (#9) was indiscriminate.

The persistence of the cats in attacking an object was measured in this same experiment. An individual trial was regarded as persistent if the cat continued to attack the object for at least two seconds after the first contact. If the total number of the attacks on a given category of objects is 100 per cent, we find that 89 per cent of the attacks made on the anesthetized rat were persistent (i.e., they continued for at least two seconds). On the stuffed rat, 67 per cent and on a dummy, 48 per cent were persistent. In short, the cats are selective, and their persistence in attacking is related to the type of object attacked.

What are the senses by which a stimulated cat locates and finally

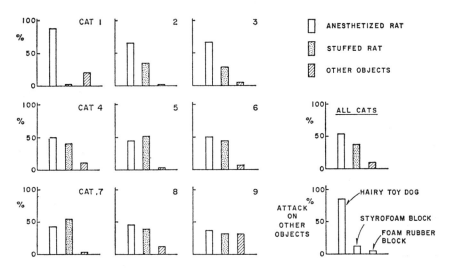

FIGURE 3 The relative frequency with which each of nine cats attacked one of three objects is shown in these graphs. The attacks were made most frequently either on the anesthetized or the stuffed rat, except in the case of Cat 9. The few attacks on the third class of objects were made primarily on the hairy toy dog.

bites a rat? In a large space (12' x 6') vision is of critical importance.[19] The frequency of attacks drops from 100 per cent to 40 per cent or less when the animal is blindfolded, as seen in Figure 4. There is a

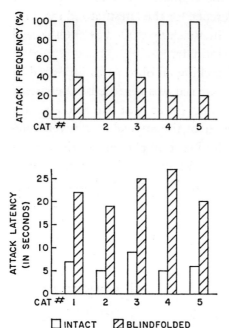

FIGURE 4 Each of five intact cats, when stimulated at sites in the hypothalamus, attacked a rat on all trials, with a mean latency of less than nine seconds. The frequency of attacks was reduced from 100 per cent to 45 per cent or less when a cat had a blindfold covering its eyes. The time from beginning of stimulation until attack occurred increased correspondingly. The enclosure in which the cats were tested measured 12' x 6'. Vision is less important in a smaller cage.

corresponding increase in the time to attack, that is, the time from the beginning of stimulation until the cat bites or strikes the rat. On the other hand, olfaction seems to play a relatively unimportant role, despite the macrosmatic character of the cat's brain. When the olfactory bulbs were removed, the incidence of attack was unchanged (Figure 5), and in five instances out of nine there were no significant increases in time to attack, although there were in the other four instances. A blindfolded cat with intact olfactory apparatus will pass within a few inches of a rat without turning to it. If a cat is both blindfolded and deprived of its olfactory bulbs, it will still attack vigorously if it touches the rat with forepaws or muzzle, just as it will when vision and olfaction are intact.

Two sensory branches of the trigeminal nerve are of critical importance for the act of biting.[19] Attack was completely blocked when the

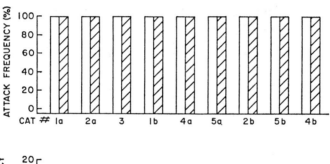

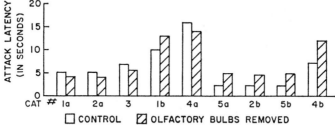

FIGURE 5 The frequency of attack was unchanged after olfactory bulbs were removed. The latency to attack was also unchanged in the five instances on the left side. In the four instances on the right side of the figure the latency was slightly, but significantly, increased. The role of olfaction in attack is surprisingly small. The cat is indicated by a number, and the stimulated site by a letter. Cat 4 had a lesion in the mesencephalon before the present comparison was made.

infraalveolar and infraorbital branches that mediate sensation around the mouth were cut. (The motor branches of the trigeminal nerve were left intact, and the cats ate without noticeable difficulty). When the animals were stimulated in the hypothalamus, they did not open their mouths and bite the rat, even though they found it successfully. They rubbed their muzzles back and forth over the rat without biting it. Figure 6 shows the results of sectioning the branches of the Vth nerve in six cats. The attack that is defined as biting or striking a rat dropped from 100 per cent to 0. Some cats were able to attack after their trigeminal nerves were cut. When they were blindfolded, they, too, were unable to open their mouths to bite during stimulation. These results are presented in Figure 7. It should also be noted that when those cats that could bite if not blindfolded did so, they

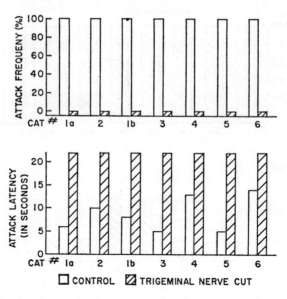

FIGURE 6 In these six instances, the frequency of attack fell from 100 per cent to zero per cent after the infraorbital and infra-alveolar branches of the trigeminal nerve were cut. The number indicates the cat, and the letter refers to an electrode.

took only one bite and not repeated bites, as would a normal animal. The visual stimulus was adequate for the first bite, but not for repetition. These data indicate that stimulation of some sites within the hypothalamus permits visual stimuli to be effective in opening the mouth, while visual stimulation is ineffective when other "attack" sites within the hypothalamus are stimulated.

The area around the mouth of cats with intact trigeminal nerves was examined with a small tactile probe during electrical stimulation of those sites in the hypothalamus that could elicit attack.[18] During stimulation of the hypothalamus, one area around the mouth produced head movement on being touched: this movement brought the object to the lips. Tactile stimulation of a second region along the lipline during stimulation of the hypothalamus brought about rapid opening of the mouth, followed by a strong bite. This reflex was absent in cats in which the trigeminal nerves had been sectioned. These regions are shown in Figure 8.

Another aspect of the response is interesting. The size of the recep-

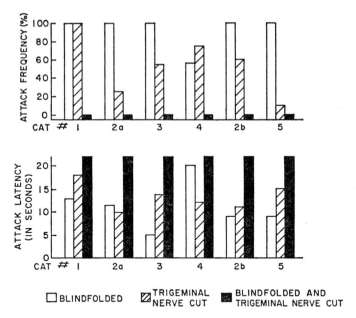

FIGURE 7 In these five instances, cutting the branches of the trigeminal nerve did not abolish attack. When these animals were also blindfolded, attack fell to zero. Blindfolding by itself did not result in as severe a deficit in this experiment as in the data in Figure 4, because these cats were tested in a smaller cage. The number indicates the cat, the letter the electrode in the cat. Olfactory bulbs were removed from Cats 2–5 before this comparison was made. A lesion in the mesencephalon of Cat 5 had also been made prior to the test.

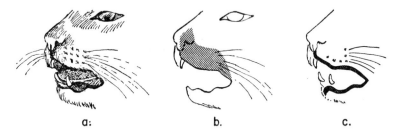

FIGURE 8 Cross-hatching (b) shows area of upper lip in which electrical stimulation produces head movement that brings a tactile stimulus to the lips. Heavy line (c) indicates area producing jaw opening when stimulated. A more complete drawing of the cat's head is in (a).

tive fields for head movement and for mouth opening increases with increasing intensity of stimulation of the hypothalamus.[18] When a small probe is moved from the corner of the mouth toward the midline at low intensities of hypothalamic stimulation, the mouth first opens when the probe reaches the region of the canine teeth. This response is elicited by touching the lipline between that region and the midline. At higher intensities, opening first occurs closer to the corner of the mouth. At still higher intensities, touching the corner is sufficient to bring about mouth opening. Data from nine separate experiments are shown in Figure 9.

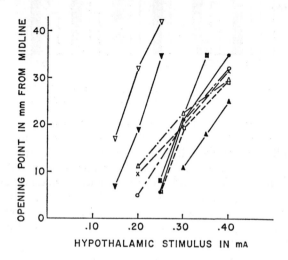

FIGURE 9 A tactile stimulus was moved from the corner of the cat's mouth toward the midline. The first point at which mouth opening occurred was measured for various intensities of stimulation of the hypothalamus. The mean distance from the midline increased with increasing stimulation to the hypothalamus. The entire region from the midline to the opening point was effective for the elicitation of jaw opening.

In addition, electrical stimulation of part of the motor nucleus of the Vth nerve causes the jaw to close. If the hypothalamus is stimulated concurrently with stimulation of the motor nucleus, the jaw closes more fully. An example of this increase is shown in Figure 10.

In brief, then, stimulated cats use three sensory systems — vision,

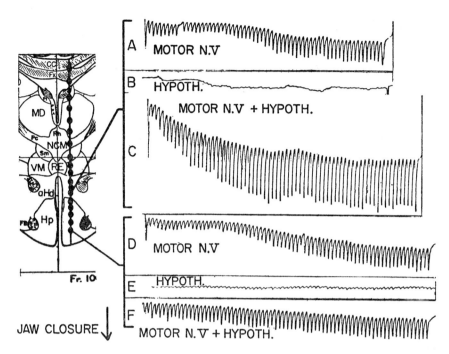

FIGURE 10 Combining stimulation of the hypothalamus with stimulation of the motor nucleus of the trigeminal nerve resulted in an increase in the closing movement of the jaw. Tracings A and B show the effects of stimulating the motor nucleus alone and the hypothalamus alone. Tracing C shows the combined effect. Tracings D, E, and F show that hypothalamic stimulation at a different point is ineffective. This effect has been studied in twelve cats.

olfaction, and touch — to varying degrees in locating a rat. Mouth opening to initiate biting depends on two sensory branches of the trigeminal nerve. These are some of the peripheral mechanisms mediating a cat's attack on a rat.

At least three categories of central mechanisms that mediate attack may be distinguished. First are sensory and motor regions, which, if stimulated directly, do not elicit attack, but are influenced by stimulation of sites that do. An example of this category has just been given. Second are effective sites (i.e., those regions of the brain which, when stimulated, bring forth attack). Third are regions within the brain which, on being stimulated directly, do not elicit attack or a competing response, but modulate the activity elicited by concurrent stimulation of effective sites, either facilitating or inhibiting the site's activity.

Figure 11 is based only on our own data of category two, which are in substantial agreement with those of earlier authors, such as Hess and Brügger,[11] Ranson,[25] and Hunsperger.[12] It differs from them primarily in terms of the emphasis on quiet biting attack. The areas for quiet biting attack in both the hypothalamus and midbrain are somewhat lateral to those from which Hess elicited affective defense. In addition, we found that parts of the thalamus, a structure not previously implicated in rage, were effective sites for eliciting attack.

The relationship of these various effective sites to one another is under study. From Bard's early work[2] it is clear that the posterior hypothalamus is required for ready elicitation of sham rage. Attack is more complex than sham rage, usually involving forebrain mechanisms, so the hypothalamus might also be required for attack. On the other hand, attack can be elicited from the midbrain, and most aspects of sham rage are elicitable in decerebrate cats on a reflex basis, even though intense stimuli are required.[33] Perhaps the midbrain and the rest of the forebrain apart from the hypothalamus are sufficient for the attack phenomenon. The hypothalamus was isolated in great part from the rest of the brain in four cats,[7] which were observed as they attacked mice or rats. Two of the animals attacked upon stimulation of the midbrain; the other two attacked when they saw a mouse. These studies show that the great part of the hypothalamus is not essential to attack, even though it may function regularly in a normal cat. The data from five additional cats with large but incomplete lesions of the hypothalamus support this conclusion.

The third category of neural mechanism may facilitate or inhibit attack without necessarily giving rise to it. The structures investigated thus far are the amygdala,[3,14,29] the hippocampus,[20,24] the midbrain reticular formation,[16,22] and the thalamus. The first two are

---

FIGURE 11 The sites at which monopolar stimulation leads to attack are presented in frontal sections of the cat's brain, according to the atlas prepared by Jasper and Ajmone-Marsan.

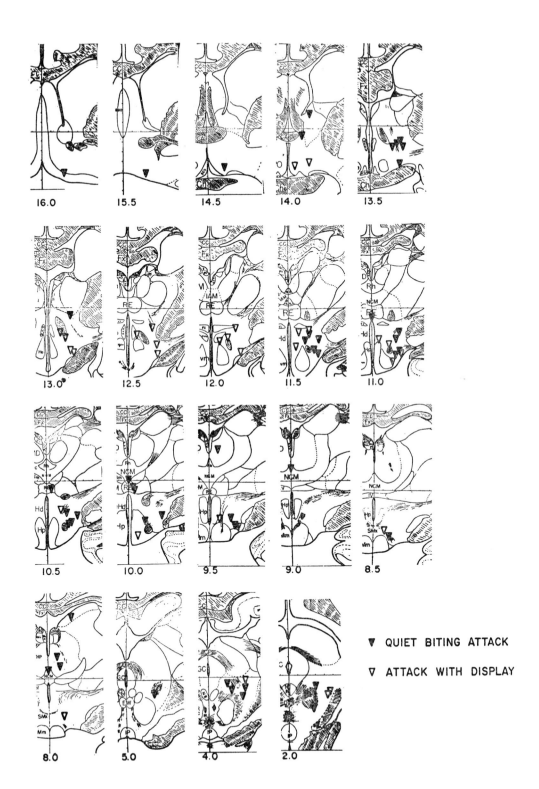

part of the limbic system, the second belongs to the reticular system, and the thalamus is connected with both.

In the investigation of the modulating structures, the basic procedure has been to stimulate a second structure concurrently with the hypothalamus. The amygdala and hippocampus are prone to afterdischarges, so the intensities employed in stimulation were below those sufficient to initiate afterdischarges. (The electrical activity of the cat's brain was monitored on all trials to ascertain the presence or absence of afterdischarges.)

Low-level stimulation of the amygdala concurrently with that of the hypothalamus had a differential effect, depending upon the amygdaloid region stimulated.[6] Stimulation of the region around the junction of the magno-cellular portion of the baso-medial nucleus with the lateral nucleus of the amygdala delayed attack, while stimulation of the dorso-lateral section of the posterior portion of the lateral nucleus facilitated it, although the facilitation effect was less pronounced than was the delaying effect. Figure 12 is a presentation of data obtained by concurrent stimulation of the amygdala.

Similar data were obtained when the hippocampus was stimulated

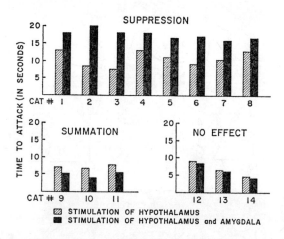

FIGURE 12 When stimulation of the amygdala was combined with stimulation of the hypothalamus, the time from beginning of stimulation until attack occurred was decreased, increased, or unchanged, depending upon the site stimulated in the amygdala. Consult text for details.

concurrently with the hypothalamus.[30] Stimulation of the dorsal hippocampus delayed attack; stimulation of a region in the ventral hippocampus, the subjacent pyriform cortex, and the posterior part of the lateral nucleus of the amygdala all led to facilitation. The results of this experiment are summarized in Figure 13. The stimulation of

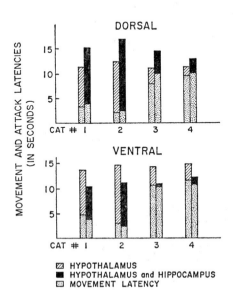

FIGURE 13 Two sites within the hippocampus and one in the hypothalamus were stimulated in combination. Stimulation of the dorsal hippocampus has consistently increased the latency to attack; stimulation of the ventral hippocampus speeded the attack. Similar results were obtained in eight cats. Stimulation of the pyriform cortex and subiculum adjacent to the ventral hippocampus gave a similar result. The time from the beginning of stimulation until the cat first made a translational movement constitutes the movement latency. The changes in attack latency are not attributable to movement latency alone.

the dorsal hippocampus produced an effect comparable to that produced by bilaterally propagated hippocampal afterdischarges.[32]

It has also been found that sites along the midline portion of the thalamus in dorsal-medial nucleus and nucleus reuniens facilitate and inhibit attack elicited by stimulation of the hypothalamus.[17] These phenomena are illustrated in Figure 14.

In stimulating the midbrain reticular formation, care was taken to avoid sites that produced reactions incompatible with attack when stimulated at high intensities. Many pathways course through the midbrain, and it is possible to stimulate some that are not related to the reticular system. Stimulation of selected sites, at intensities sufficient to produce only mild alerting, elicited facilitation of attack.[28] These results are presented in Figure 15.

Figure 16 summarizes the neural mechanisms involved in rage and

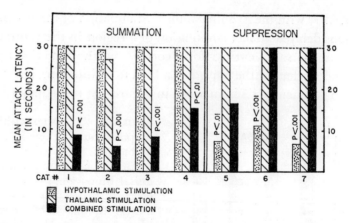

FIGURE 14 Stimulation of sites within the thalamus also summates with or counteracts stimulation of the hypothalamus. The sites are close to those illustrated in Figure 11.

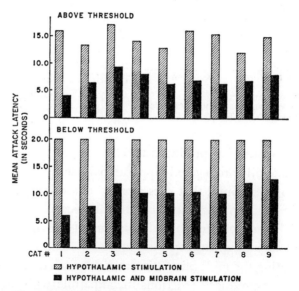

FIGURE 15 Stimulation of the midbrain, when combined with stimulation of the hypothalamus (at two different intensities), resulted in a reduction of the time to attack.

attack. The patterning mechanism represents neural mechanisms from which attack can be elicited when activated at sites in the hypothalamus, midbrain, and thalamus. Facilitating and inhibiting mech-

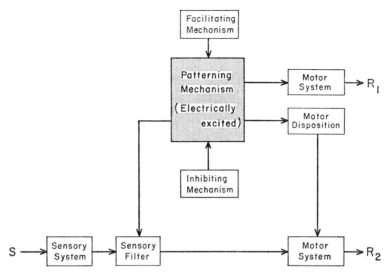

FIGURE 16 Summary diagram. See text.

anisms are depicted as acting upon the patterning mechanism. These modulating sites are the appropriate parts of the amygdala, hippocampus, and thalamus.

The effect of the patterning mechanism on the motor system is twofold. First is a direct effect that accounts for motor responses (R_1) that appear to be relatively independent of environmental conditions: examples are walking, sniffing, snarling, and the display of rage. Second is a motor disposition. If an adequate stimulus is present, a second response labeled R_2 appears. Examples are attack on a rat and the effect of hypothalamic stimulation on jaw closure. If one stimulates the motor nucleus for jaw closure and obtains a response of a given size, the response will be found to increase with concurrent stimulation of a site in the hypothalamus from which attack can be elicited. Stimulation of the hypothalamus facilitates jaw closure at the level of the motor nucleus. This kind of motor disposition is shown in the diagram.

We have also indicated that the patterning mechanism has an effect upon the sensory systems, although we cannot as yet state the level at which this effect occurs. We have demonstrated that the lip region, which constitutes the effective receptive field for mouth opening, functions according to the intensity at which the hypothalamus is stimulated. We have also found that inside the cat's mouth a tactile stimulus produces closure during hypothalamic stimulation. Similarly, visual stimuli are adequate to produce jaw opening when certain attack-eliciting hypothalamic electrode sites — but not others — are stimulated; this indicates that stimulation influences the effectiveness of sensory inputs in eliciting a reflex response. We have also found that a cat stimulated at one hypothalamic site will attack an experimenter, leaving a rat untouched, and if stimulated at another hypothalamic site, will attack the rat savagely, and ignore the experimenter. All this evidence points to a sensory gate controlled by stimulation of effective sites within the brain. We have assigned these functions to the sensory filter.

In brief, stimulation of the hypothalamus, whose activity can be modulated from other sites in the brain, sets up motor activities and motor dispositions. These last emerge as responses when triggered by stimuli that meet the requirements established by neural mechanisms activated in the hypothalamus.

Brain Mechanisms and Emotion

RONALD MELZACK

I would like to begin this discussion by telling you about my daughter. She is five years old, plump and cute, and she loves to be tickled — by me or my wife, but not by strangers. She often asks me to tickle her, which I do gladly, and she laughs uproariously, squirms around in delight, and asks for more when I stop. If I tickle too hard, or change my expression and pretend to scowl, she tries to get away from

---

RONALD MELZACK Department of Psychology, McGill University, Montreal, Canada

me. If I persist, she becomes frantic in her attempts to escape, and her laughter may turn to tears. The moment my tickling movements become gentle again, or my scowl changes back to a smile, all is well, and the game continues as before. This description also applies to my four-year-old son, who is even more ticklish than my daughter. Sometimes it is not even necessary for me to touch him: all I have to do is extend my hand and *say* that I am going to tickle him, and that is enough to get him laughing.

These events, I believe, contain in a nutshell a description of three salient features of emotion. 1) There is an emotion-provoking *sensory input* — in this case, repetitive, light touches applied to particular body areas in the context of play or games. 2) A *high level of arousal* or excitement pervades the whole behavior pattern — but it is not simply arousal, because it is clear that there is a strong affective coloring to the experience[27] that is determined partly by the intensity of the tickling movements. At low intensity, the input is pleasurable but at a higher intensity it may become "too much," and behavior shifts from tickle-seeking to escape, indicating a switch in both affect and drive states. 3) *Central mediating processes* — such as my daughter's interpretation (based on experience and imagination) of my scowl or of the intentions of a stranger — play an astonishingly powerful role in determining whether the tickle (with input intensity held constant) is very pleasant or very unpleasant.

Although Dr. Pribram did not address himself to the problem of tickle, he nevertheless described these three features of emotion, in all their awesome complexity. Emotion was once so simple! It was neatly compartmentalized as a separate and distinct psychological entity, certainly deserving at least one major brain center. Conceptually, Gall and Spurzheim's (cited by Boring[1]) phrenological maps (Figure 1) characterize these notions by attributing a cortical center (which has a corresponding skull bump) to each emotion (or "affective faculty"). Although phrenological maps disappeared long ago, they left their mark on psychological concepts, and investigators have repeatedly sought one or more centers for the emotions. However, as Dr. Pribram has noted, modern psychology and physiology make it virtually impossible to separate "emotion" from motivation

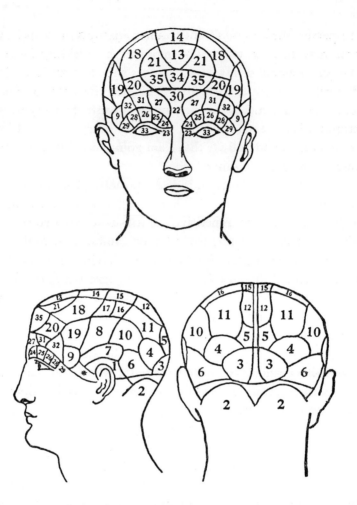

FIGURE 1 The "powers and organs of the mind," according to Spurzheim, *Phrenology, or the doctrine of mental phenomena,* 1834. The "affective faculties" are separated into "propensities" and "sentiments." The "propensities" are: ? — desire to live; * — alimentiveness; 1 — destructiveness; 2 — amativeness; 3 — philoprogenitiveness; 4 — adhesiveness; 5 — inhabitiveness; 6 — combativeness; 7 — secretiveness; 8 — acquisitiveness; 9 — constructiveness. The "sentiments" are: 10 — cautiousness; 11 — approbativeness; 12 — self-esteem; 13 — benevolence; 14 — reverence; 15 — firmness; 16 — conscientiousness; 17 — hope; 18 — marvelousness; 19 — ideality; 20 — mirthfulness; 21 — imitation. The remaining "bumps" represent "intellectual faculties." (From Boring.[1] Reprinted by permission of Appleton-Century-Crofts)

and learning, memory and imagination, thought and perception. Emotion is influenced by all these activities and contributes to them. Widespread portions of the nervous system are clearly involved in emotion, yet each portion has its specialized functions. Dr. Pribram has dealt with what these functions are, and how they are interrelated.

The history of the problem of emotion, as Dr. Pribram described it, is a fascinating progression of ideas. From the peripheralist theory of James and Lange, which places heavy emphasis on visceral activities and the autonomic nervous system, we moved to the more centralist notions of Cannon and Bard and of Papez and MacLean, in which brain structures—the thalamus and the limbic system—play an increasingly prominent role. Dr. Pribram, in his present theory, has emphasized the role of memory and experience in emotion in which the cerebral cortex is involved, and has therefore brought us to a yet more centralist approach. He has noted that the remembered outcome of previous actions, and our experience with similar configurations of events, are critical in determining our reactions in emotional situations. Emotion, then, once considered to be a holdover from our animal ancestry and unbecoming the mature human adult (excepting, perhaps, filial love and patriotism) is now seen to have a major cortical contribution.

Mediating processes such as thoughts and past experience, however, may not only influence emotion; they may evoke it. The *thought* of war, of betrayal by a friend, of one's children, may all evoke strong emotion—anxiety, hate, or love. Moreover, as Hebb[9] has pointed out, mediating processes hold the key to distinguishing such emotions as fear and anger. Both may involve the same intensity of arousal, yet both are subjectively and behaviorally different, and that difference lies, at least in part, in the mediating processes that accompany each. The tendency to flee or to attack may be determined by past experience in similar situations, by the provoking agent being your boss or your wife, and so on.

This emphasis on mediating processes does not, of course, deny the importance of sensory events. Indeed, Dr. Pribram has placed sensory mechanisms at the center of his theory. He has described a complex model of emotion in which orienting responses to novel sensory inputs (or inputs incongruous with past experience) are equated with "perturbation" and "disequilibrium." There is, to be sure, a class of novel, strange, unusual objects that produces irrational or spontaneous fears.[7] These objects are characteristically incongruous with past experience: an umbrella that opens suddenly produces fright in a dog;[14] a death mask of a chimpanzee elicits fear in a chimp;[7] the sight of a mutilated animal or person evokes horror in the human.[9] In general, however, novel stimulation and orienting responses cannot be equated with perturbation and emotion. Some novel inputs elicit exploration and curiosity.[8] Other forms impinge on us daily without any emotional impact: a "mutilated" chair rarely evokes fear. Novelty, then, does not always produce emotion. Conversely, all emotions cannot be ascribed to novel sensory inputs.[16] There is a gap between the sensory input and emotion in Dr. Pribram's model that still needs to be bridged.

An important clue to the role of the sensory input in evoking the affective dimension of emotion lies in the convincing evidence that many sensory inputs are desirable up to some intensity level (that is, are accompanied by positive affect), but that beyond this critical intensity they become less desirable or even aversive. There are no better examples of this than Pfaffmann's experiments with different concentrations of salt and sugar solutions. Figure 2, from one of his experiments, shows that animals make more and more approaches to a solution as the concentration increases. But beyond the critical concentration level they approach less often and finally avoid it. Records from peripheral nerve show increased firing rates proportional to the solution concentration. Similarly, rats work to get moderate levels of light stimulation—the brighter the light the harder they work[24]—but actively avoid intense light.[10] Intensity, then, plays an important role in affect and drive.[21,27] When the input, after it has undergone modulation by central mediating processes, exceeds some critical intensity level, motivational tendency shifts from approach to avoidance and, in terms of human experience, affect switches from positive to negative, from pleasantness to unpleasantness.

Perhaps the most characteristic feature of emotion is the high level of activation or arousal. Lindsley,[12] Hebb,[8] and others have stressed

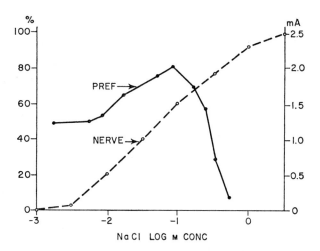

FIGURE 2 Preference response to different salt concentrations and magnitude of the chorda tympani nerve discharge in the rat. Ordinate at the left corresponds to preference response, ordinate at the right shows magnitude of integrator deflection. (From Pfaffmann.[20] Reprinted by permission of the author and Reinhold Publishing Co.)

the critical role of arousal (and the reticular activating system), and Dr. Pribram has here reaffirmed it, at the same time pointing out (and I agree with him) that arousal by itself must not be equated with emotion. Arousal is a vague term, although we generally know what we mean by it psychologically. Physiologically, of course, it is identified with EEG activation, autonomic nervous system activity, and so forth. Psychologically, arousal seems to be more than this. There is, I believe, an affective component to arousal. High levels of arousal in the human are almost always accompanied by positive or negative affect. Excitement and interest, up to a certain level, are all moderately high arousal states with positive affective coloring. Nervousness and anxiety are high arousal states with negative affective coloring.

My interest in the problem of pain[15] has led to a consideration of emotion because pain has long been held by some investigators to be an emotion — the opposite of pleasure — rather than a sensation. Indeed, this idea of pain is part of an intriguing and usually neglected bit of history.[2] At the turn of the century, von Frey and Goldscheider fought a bitter battle on the question of pain specificity.[26] Von Frey argued that there *are* specific pain receptors, while Goldscheider contended that excessive stimulation and central summation are the crucial determinants of pain. But there was a third man in the battle — H. R. Marshall,[13] an early philosopher and psychologist — who said, essentially, "a plague on both your houses; pain is an emotional quality, or *quale*, that cuts across all sensory events." He admitted the existence of a pricking-cutting sense, but pain, he thought, was distinctly different. All sensory inputs, as well as thoughts, could have a painful dimension to them, and he talked of the pain of bereavement, the pain of listening to badly played music.

His extreme approach was, of course, open to criticism. Strong[25] and Sherrington,[23] for example, noted that the pain of a scalded hand is different from the "pain" evoked in a musicologist by even the most horrible discord. Marshall was soon pushed off the stage. But if a less extreme view is taken of his concept, it suggests a new approach to pain. For pain does not have just a sensory dimension; it also has a strong negative affective dimension, and it drives us into activity. We must do something about it — avoid it, attack the source, and so on — and, of course, these response patterns are in the realm of emotion and motivation.

Kenneth Casey and I[17] are presently working with a theoretical model in the attempt to account for the emotional and motivational dimension of pain in addition to its more obvious sensory dimension. I would like to outline these ideas briefly, and then try to relate some of them with those expressed by Dr. Pribram. Not long ago, Patrick Wall and I proposed a Gate Control theory of pain (Figure 3).[18] We suggested that in the dorsal horns there exists a Gate Control System, which determines the amount of input that is transmitted from the peripheral fibers to dorsal horn transmission (T) cells. Moreover, we proposed that brain activities, such as attention and memories of prior experience, are also able to influence the gate, so that the number of impulses transmitted per unit time by the T cells is determined by the ratio of large- and small-fiber inputs, and by the downflow from the brain. (This feedback loop through the brain could well perform the match-mismatch functions that Dr. Pribram has described

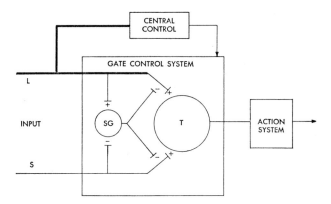

FIGURE 3 Schematic diagram of the Gate Control theory of pain mechanisms. L, the large-diameter fibers; S, the small-diameter fibers. The fibers project to the substantia galatinosa (SG) and first central transmission (T) cells. The inhibitory effect exerted by SG on the afferent fiber terminals is increased by activity in L fibers and decreased by activity in S fibers. The central control trigger is represented by a line running from the large-fiber system to the central control mechanisms; these mechanisms, in turn, project back to the Gate Control System. The T cells project to the entry cells of the Action System. +, excitation; —, inhibition. (From Melzack and Wall[18])

to us.) Wall and I noted that pain experience and response involve a complex sequence of activities: orienting, vocalization, autonomic responses, avoidance or aggression, and so forth. These, we proposed, are the function of the Action System, which clearly comprises widespread portions of the brain. We suggested that the signal which triggers the Action System occurs when the output of the T cell reaches or exceeds a critical level.

Casey and I have been considering the activities of that output at the brain level, and this (Figure 4) is the simplest model we have been able to come up with. Three major categories of activity appear to be involved in pain processes. The first is the selection and modulation of the sensory input during transmission to provide the neural basis for the sensory-discriminative dimension of pain. These activities subserve the perception of sensory qualities that reflect the location, magnitude, and spatio-temporal properties of the input, and which may be completely devoid of negative affect.

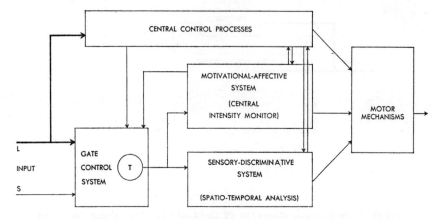

FIGURE 4 Schematic diagram showing the three categories of activity involved in pain processes at the brain level: (1) central control processes, (2) activity of the motivational-affective system which comprises the "central intensity monitor," and (3) spatio-temporal analysis of the input. The three "blocks" of activity interact with one another and with the Gate Control System, and their output impinges on motor mechanisms. (From Melzack and Casey[17])

The second category is related to affect and motivational tendency, and we have labeled the underlying mechanisms as the "central intensity monitor." The activity in this system subserves the negative affect and aversive drive that are essential dimensions of the whole pain experience. These two dimensions are brought clearly into focus by clinical studies. Patients who have undergone frontal lobotomy say that they have pain but it does not bother them.[5] The affective component of pain is strikingly diminished while the sensory features are relatively undisturbed; indeed, sensory thresholds may be lowered.[11] Descriptions of "pain asymbolia"[22] and the cases of "congenital insensitivity to pain" described by Ford and Wilkins[4] also suggest a separation of the sensory and affective dimensions of pain.

The third category in Figure 4 comprises the mediating or central control processes that contribute to pain: experience, attention, expectation, and so forth.[15]

We presume that these three "blocks" of activity interact with one another and with the Gate Control System, and that their outputs impinge on motor mechanisms to provide: 1) perceptual information regarding the location, size, and temporal-spatial properties of the noxious stimulus; 2) motivational tendency toward fight or flight; and 3) information based on prior experience, probability of outcome of different response strategies, and so on.

Casey and I[17] have reviewed the physiological and anatomical evidence suggesting that portions of the reticular and limbic systems may carry out the functions of the central intensity monitor. Cells in the reticular formation are capable of both spatial and temporal summation of somato-sensory inputs from the dorsal horn T cells, so that discrete spatio-temporal information is transformed into intensity information. We propose that the output of these reticular cells, up to a critical intensity level, activates those areas subserving positive affect and approach tendencies.[19] Beyond that level, the output activates other areas underlying negative affect and aversive drive.[3] We suggest, therefore, that the drive mechanisms associated with pain are activated whenever the input into the motivational-affective system exceeds the critical level. This notion fits well with Grastyan's[6] observations that animals seek low-intensity electrical stimulation of some limbic system structures, but avoid or actively try to stop high-intensity stimulation of the same areas. These signals from the limbic system to motor mechanisms set the stage for avoidance or aggression by whatever means is most effective on the basis of both central control processes (such as prior experience) and the sensory information.

Here, then, is another way of saying some of the things that Dr. Pribram was proposing. The model is explicit (perhaps too explicit), because it deals with a single experience rather than with the whole gamut of emotions. But it sets the stage for consideration of Dr. Pribram's ideas of the salient features of emotion: 1) the input from the sensory systems and the viscera, and the orienting behavior it evokes; 2) the arousal or activation, which is clearly a dimension of emotion, but which, as Dr. Pribram has shown effectively, cannot be the whole answer to the problem of emotion; and 3) cognitive processes in the form of memory, against which the input is matched, and which provides the direction and course of emotional experience and response.

Emotion and the Sensitivity of Psychoendocrine Systems

JOSEPH V. BRADY

DISCUSSION

The Conditions for Emotional Behavior
GEORGE MANDLER   96

Psychoendocrine Systems and Emotion: Biological Aspects
SEYMOUR S. KETY   103

Involvement of the Hormone Serotonin in Emotion and Mind
D. WAYNE WOOLLEY   108

As Charles Sherrington observed long ago, the problem of emotion constitutes one of the most fruitful meeting grounds for the combined disciplinary skills of psychology and physiology. Indeed, the subject matter that provides the theme for this conference has long occupied the attention of biological and social scientists of all persuasions concerned with the general problems of behavior and the related events of an organism's physiology. The experimental analysis of relationships between those organismic-environmental interactions considered emotional and the neural-physiological participants in such behavioral events has proceeded slowly and somewhat haltingly, however, amidst a host of biological and social-psychological complexities. Introspective emphasis upon the phenomenological aspects of the emotion problem has traditionally occupied a far more prominent place in psychological theorizing than has objective scientific descrip-

---

JOSEPH V. BRADY Walter Reed Army Institute of Research and University of Maryland

tion. Refinements in methodology only recently have permitted physiological measurements appropriate to the interdisciplinary analysis of such complex psychophysiological relationships.

Over the past decade, the technological advances emerging from the analysis of behavior in laboratory animals and the development of reliable microanalytical methods for biochemical measurement of hormones in blood and urine have made possible relatively direct experimental study of previously unexplored facets of psychophysiology. Within the framework of this rapidly developing psychoendocrine approach, emotion has received appropriate emphasis in much initial investigative activity, and the problems and progress, both conceptual and methodological, associated with such an interdisciplinary experimental analysis provide the primary focus for this report.*

In probably no other domain of behavioral science has so little empirical data provided the occasion for so much theoretical speculation as has the general area of emotion. Indeed, the very word emotion has been the object of persistent reification in the vain search for imaginary causes for behavior and the fruitless perpetuation of explanatory fictions. It seems unnecessary to dwell at length upon the multitude of semantic, linguistic, and taxonomic difficulties that have produced this wide gap between scientific operations and systematic interpretation, although one continuing source of confusion would seem to require at least brief comment before we consider the interrelations between endocrine systems and emotion.

Since at least the time of James and Cannon, the interchangeable use of terms referring to "feelings" or "affects" on the one hand, and "emotional behavior" on the other has encouraged endless polemic exchange, occasioned primarily by a failure to differentiate between two distinguishable classes of psychological events. Although both feelings and emotional behavior involve psychological interactions between organism and environment, a useful and important distinc-

---

* Most of the experimental work to be described is the result of a collaborative research effort with Dr. John Mason, Chief of the Department of Neuroendocrinology at the Walter Reed Army Institute of Research and other staff members of the Division of Neuropsychiatry, as indicated in the references to previous publications. Responsibility for the theoretical and interpretive portions of the presentation, however, must be borne solely by the author.

tion between the two would seem possible on the basis of the localizability of their principal effects or consequences. On the one hand, emotional behavior would seem to be most usefully considered as part of a broad class of *effective* behaviors primarily directed toward changing some aspect of the organism's exteroceptive stimulus environment. By contrast, feelings or *affective* behaviors would seem to be distinguishable as a generic class of interactions between organism and environment, the primary consequences of which are localizable *within* the reacting organism rather than in exteroceptive stimulus objects. Many different subclasses of feelings may be identified within this broad affective category, but emotional behavior seems most parsimoniously considered to be only one unique class clearly separable from many other classes of affective and effective interaction patterns.

The relevance of this differentiation between feelings and emotional behavior to psychophysiological research on emotion can perhaps be seen more clearly in considering some of the defining properties of these two classes of behavioral events. Perhaps the most salient characteristic of feelings is their intimate association with autonomic-visceral, endocrine, and proprioceptive activities, which have provided a recent focus for numerous conditioning studies documenting discriminative control involving such interoceptive stimulus events.[17-19] The ease with which these visceral, endocrinological, and proprioceptive responses can be transferred from one object or situation to another emphasizes the predominantly conditioned Pavlovian nature of eliciting stimuli for feelings, or affective reactions. And the flexibility in organization that characterizes the loose integration of feeling responses with exteroceptive environmental stimulus objects can be best understood in terms of the internal focus of their reactive consequences. Under such conditions, the proliferation and variability of feeling responses appear to be limited only by the intricacies of an organism's past conditioning history and the complexity of environmental stimulus situations.

Understandably, the labeling of such predominantly private feeling events can be expected to present many problems for the development of an appropriate verbal repertoire, and the widespread difficulty traditionally associated with affective communication provides ample testimony to the operational inadequacy of available vocabularies. To the extent that the participation of interoceptive processes in behavioral interactions represents a necessary condition for the occurrence of feelings, a functional experimental analysis emphasizing such psychophysiological interrelationships may potentially contribute at least some definitional clarity in an approach to the broad problem of emotion.

Indeed, the very prominence of such visceral-endocrine participation in at least some *effective* response patterns has doubtless contributed to the persistent confusion between feelings and emotional behavior. In accordance with the present formulation, however, such psychophysiological activity is by no means a necessary condition for emotional behavior defined in terms of the primarily external localization of effects with respect to the reacting organism, and involving changes or perturbations, characteristically episodic, in the ongoing course of organismic-environmental interactions. Viewed in this way, such emotional changes in ongoing behavior become susceptible to a functional analysis of their conditions of occurrence in relation to explicitly defined experimental operations, principally the application of both conditioned (e.g., Pavlovian) and unconditioned (e.g., food, shock) reinforcers. The experimental analysis of the special character of these perturbations in the ongoing interaction between organism and environment, specifically as they relate to changes in the reinforcing value of different primary reinforcers, would seem to provide an appropriate starting point for a rigorous and systematic treatment of this aspect of the emotion problem.

However great the difference between affective and effective behavior patterns, it should, of course, be obvious that feelings and emotional behavior may be operationally interrelated. Complex behavior situations frequently involve close temporal associations between the two, although their appearance need not always be simultaneous. Clearly, the internal consequences of prior-occurring *affective* feeling interactions may influence an organism's disposition to engage in certain *effective* emotional behavior. The angry feelings produced by an insulting remark frequently predispose an ongoing interaction between organism and environment to disruption by aggressive or hostile emotional behavior. Similarly, an abrupt, often chaotic, interruption of ongoing activity may well be followed with a discriminable delay, by feeling responses (e.g., fear) emphasizing the internal consequences of affective interactions that occur as sequelae to such emotional behavior. Despite this commonly observed temporal association between affective feeling responses and effective emotional behavior patterns, however, an experimental analysis of emotion would seem to require at least initial emphasis upon the distinguishable intra- and extra-organismic consequences of independently defined psychological and physiological operations.

The intimate relationship between endocrine systems and emotion has long been recognized in both the laboratory and the clinic. Over a half-century ago, the classical experiments of Cannon[3] convincingly demonstrated endocrine participation in behavioral interactions, even though some 40 years were to elapse before the introduction of reliable chemical methods for direct measurement of such hormonal changes. Only within the past two decades have developments emerging from the experimental analysis of behavior provided appropriately refined laboratory techniques for the definition and measurement of psychological processes in relationship to such physiological operations. Indeed, numerous studies have confirmed the extreme sensitivity of psychoendocrine systems to a wide range of environmental manipulations, with emotion and pituitary-adrenal cortical activity receiving appropriate emphasis.[8,10,16] Initially, experiments in this area with the rhesus monkey focused upon changes in plasma and urinary levels of 17-hydroxycorticosteroids (17-OH-CS) related to a variety of behavioral conditioning procedures involving both affective and effective interaction patterns. More recently, however, plasma and urinary epinephrine and norepinephrine measurements have provided additional indexes of sympathetic-adrenal medullary participation in such behavioral events.

The basic laboratory setting in which most of these experiments have been conducted is illustrated in Figure 1, and has been described elsewhere in considerable detail.[7] Briefly, the primate restraining chair situation provides for automatic and programable delivery

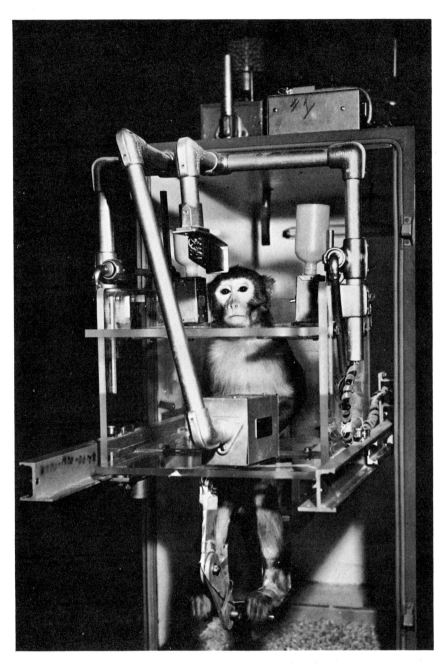

FIGURE 1 Rhesus monkey in primate restraining chair. (From Brady[1])

of food and water, administration of mildly punishing electric shock to the feet, a hand-operated electromechanical lever switch, and presentation of a variety of visual and auditory stimuli to the experimental animal. Blood is obtained from the leg of the monkey or through a chronically indwelling right atrial catheter. Urine samples are collected in a receptacle attached below the seat of the chair. Programming and control of all behavioral procedures is accomplished remotely and automatically with an electromechanical system of relays, timers, counters, and recorders.

Preliminary studies readily established that neither restraint in the chair following an initial 48-hour adaptation period, nor performance of lever pressing for food reward alone on several different schedules of reinforcement produced any significant hormonal changes in monkeys maintained for prolonged intervals under such conditions. When, however, emotional behavior conditioning procedures involving electric shock to the feet were superimposed upon such performance baselines, marked elevations in plasma 17-OH-CS levels were observed. The basic procedure producing this adrenal-cortical response is a modification of the Estes-Skinner conditioned suppression technique,[4,5] which provides a convenient laboratory model for emotional behavior within the definitional framework described above. Conditioning trials, consisting of five-minute continuous presentations of an auditory warning stimulus terminated contiguously with a brief electric shock to the feet, are superimposed upon the lever-pressing performance for food on a variable interval schedule of reinforcement. Within a few trials, virtually complete suppression of the lever-pressing behavior occurs in response to presentation of the clicker, as illustrated in Figure 2, accompanied by piloerection, locomotor agitation, and frequently urination and/or defecation.

The development of this conditioned "anxiety" response has been studied in relationship to changes in plasma 17-OH-CS levels that occurred during a series of acquisition trials. These consisted of 30-minute lever-pressing sessions, with auditory stimulus and shock pairing occurring once during each session approximately 15 minutes after the start. Seven such conditioning trials were accompanied by the withdrawal of blood samples immediately before and immediately

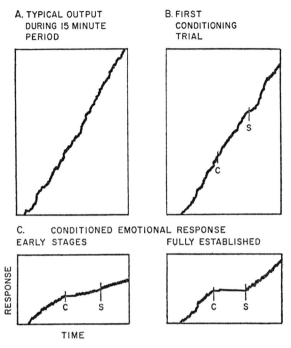

FIGURE 2 The conditioned emotional behavior as it appears typically in the cumulative response curve. (From Hunt and Brady[5])

after each 30-minute session, and 17-OH-CS levels associated with successive stages in the acquisition of the conditioned emotional behavior were determined. Figure 3 shows the corresponding changes in lever pressing and 17-OH-CS throughout the series of seven conditioning sessions. The progressive *suppression* of lever pressing in response to presentation of the auditory stimulus during each successive trial is represented by the solid lower line in Figure 3 in terms of an "inflection ratio," which provides a quantitative measure of the conditioned emotional behavior.* The broken upper line in Figure 3

---

* The "inflection ratio" is derived from the formula B–A/A, in which A represents the number of lever responses emitted during the five minutes immediately preceding introduction of the auditory stimulus, and B represents the number of lever responses emitted during the five-minute presentation of the auditory stimulus. The algebraic sign of the ratio indicates whether output increased

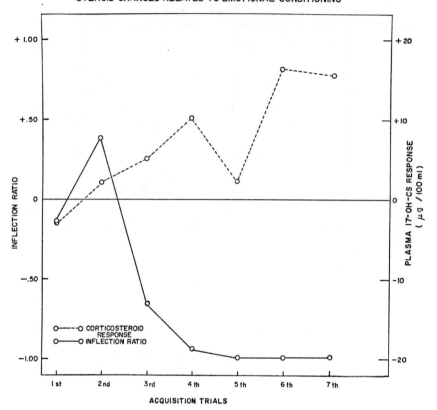

FIGURE 3 Changes in plasma 17-OH-CS levels related to emotional conditioning. (From Mason, Brady, and Tolson[13])

reflects the progressive increase in 17-OH-CS elevations during each of the seven successive "anxiety" conditioning sessions.

(plus) or decreased (minus) during the auditory stimulus, relative to the output during the immediately preceding five-minute interval. The numerical value of the ratio indicates the amount of increase or decrease in output as a fraction (percentage in decimal form) of the output prior to introduction of the auditory stimulus. Complete cessation of lever pressing during the auditory stimulus yields a ratio of −1.00, and a 100 per cent increase, a value of +1.00. A record showing essentially unchanged output obtains a ratio in the neighborhood of zero. The ratio indicates whether introduction of the conditioned stimulus produced an inflection in the output curve, how much of an inflection it produced, and in which direction. (See Hunt, Jernberg and Brady[6])

This relationship between emotional behavior and the activity of the pituitary-adrenal cortical system has been further confirmed in a series of experiments with monkeys, in which the conditioned suppression of lever pressing had been previously established. Five such animals were studied during one-hour lever-pressing sessions for food reward during alternating five-minute periods of auditory stimulus presentation and no auditory stimulus, as illustrated in Figure 4.

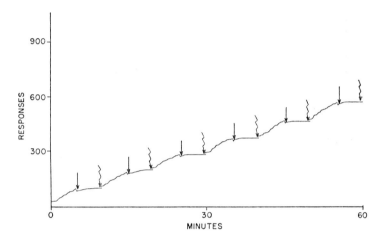

FIGURE 4 Cumulative record of lever pressing with superimposed conditioned "anxiety" response in monkey. The straight arrows indicate the onset, and the jagged arrows the termination of each five-minute clicker period. Between clicker periods the lever-pressing response rate is maintained. During clicker presentations, lever pressing is suppressed. (From Mason, Brady, and Sidman[12])

Blood samples taken before and after several such experiments with each animal, during which *no shock* followed any of the auditory stimulus presentations, revealed substantial corticosteroid elevations related to the *conditioned* emotional behavior alone. Figure 5 shows that the rate of rise of this behaviorally induced steroid elevation is strikingly similar to that observed following administration of large ACTH doses in these animals, although such pituitary-adrenal stimulation appears to cease shortly after termination of the emotional interaction, and hormonal levels return to normal within an hour. Significantly, when the conditioned anxiety response is markedly

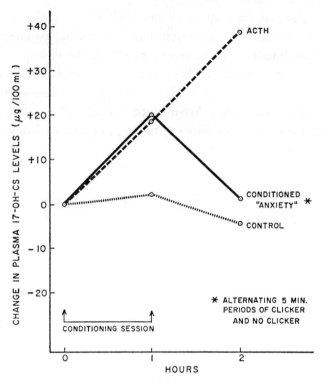

FIGURE 5 Plasma 17-OH-CS response in monkeys during conditioned "anxiety" sessions as compared to control sessions and I.V. ACTH (16 mg/kg) injection. (From Mason, Brady, and Sidman[12])

attenuated by repeated doses of reserpine administered 20 to 22 hours before experimental sessions, the elevation of 17-OH-CS in response to the auditory stimulus is also eliminated.[9]

When measurements of plasma epinephrine and norepinephrine levels were added to the corticosteroid determinations in experiments with this conditioned emotional-behavior model, the potential contributions of a "hormone pattern" approach to such psychophysiological analyses became evident. Preliminary observations in the course of a rather rudimentary conditioning experiment with monkeys, involving a loud truck horn and electric foot shock, suggested the differential participation of adrenal medullary systems in conditioned and unconditioned aspects of these emotional behavior patterns. Figure 6 shows, for example, that exposure only to the horn or

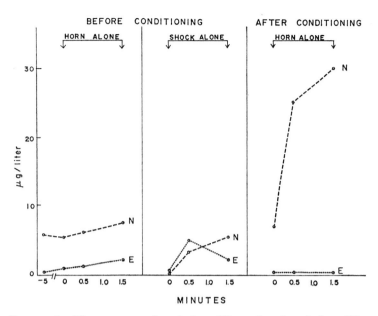

FIGURE 6 Plasma norepinephrine (N) and epinephrine (E) responses in monkey before and after emotional conditioning. (From Mason, Brady, and Tolson[13])

only to the shock *prior* to the conditioned pairing of the two, produced only mild elevations in catecholamine levels. Following a series of conditioning trials, however, during which horn sounding for three minutes was terminated contiguously with shock, presentation of the horn alone markedly increased norepinephrine levels without eliciting an epinephrine response. This hormone pattern approach has been extended in a series of experiments in which concurrent plasma epinephrine, norepinephrine, and 17-OH-CS levels were determined during monkey performance on the alternating five-minute "on," five-minute "off" conditioned anxiety response procedure illustrated in Figure 4. The results, summarized in Figure 7, were obtained during 30-minute control and experimental sessions involving recurrent emotional behavior segments, and confirm the differential hormone response pattern, which is characterized by marked elevations in both 17-OH-CS and norepinephrine, but little or no change in epinephrine levels.

Recently, it has been possible to make some preliminary observa-

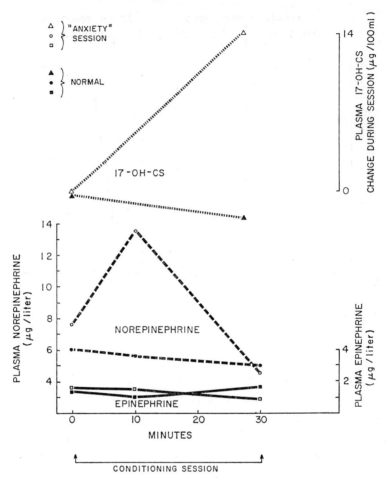

FIGURE 7 Mean plasma 17-OH-CS, norepinephrine, and epinephrine levels in the monkey during conditioned "anxiety" sessions. (From Mason, Mangan, Brady, Conrad, and Rioch[15])

tions of autonomic activity related to the same conditioned emotional behavior model. Systolic and diastolic blood pressure, as well as heart rate, have been recorded in monkeys with arterial catheters during experimental sessions, which have included both lever pressing alone and exposure to the conditioned anxiety procedure. Figure 8 shows the lever-pressing performance, heart rate, and blood pressure values obtained during approximately nine minutes of a one-hour control session prior to emotional behavior conditioning. The stable lever-

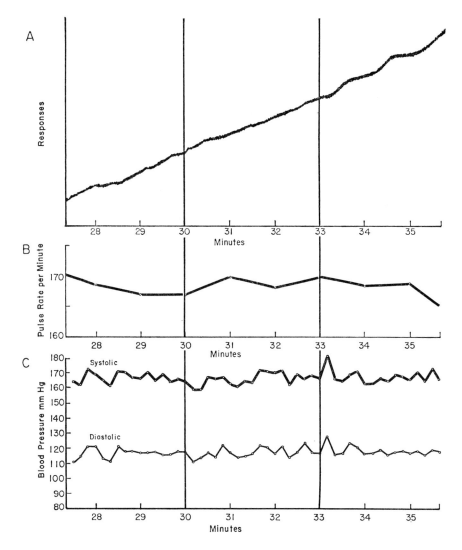

FIGURE 8 Lever-pressing performance, heart rate, and a blood pressure values during control session prior to emotional conditioning.

pressing performance was accompanied by equally stable heart rate (165 to 170 beats per minute) and blood pressure (160 to 170 mm mercury (Hg) systolic, 115 to 120 mm Hg diastolic) values throughout the session.

In contrast, Figure 9 shows the results obtained during an experimental session following a series of conditioning trials in which three-minute presentations of a clicking noise terminated contiguously with foot shock and were superimposed upon the lever-pressing performance. Complete suppression of lever pressing during clicker presentation (Figure 9A) is accompanied by a dramatic drop in heart rate from 155 to 165 beats per minute to approximately 130 beats per minute, followed by rapid recovery during the postclicker period (Figure 9B). In addition, systolic blood pressure can be seen to have fallen from approximately 160 to 130 mm Hg during the clicker, and diastolic levels, although responding less vigorously, were consistently depressed from approximately 110 to 105 mm Hg during clicker presentation (Figure 9C). By comparison with heart-rate values, however, blood-pressure levels required somewhat longer to return to baseline following termination of the clicker. Significantly, attenuation of the monkey's conditioned anxiety response, caused by a series of electroconvulsive shock treatments, is accompanied by a marked attenuation of the cardiovascular concomitants of the emotional behavior, although some residual effects of the heart-rate response to presentation to the clicker have been seen to survive even extended exposure to such ECS treatments.

Studies of a closely related discriminated "punishment" or conditioned "conflict" procedure with rhesus monkeys have further extended the analysis of endocrine patterns in relation to emotional behavior. In contrast to the conditioned anxiety procedure, which included only superimposition of the clicker-shock pairings upon lever-pressing performance, the punishment or conflict situation provided for programming both the delivery of food and the administration of shock as concurrent consequences of lever pressing in the presence of an auditory stimulus. Under such conditions, the deprived animal pressing a lever for food pellets was recurrently presented with a tone signal; at the same time, the lever responses to produce food delivered shock to the feet. Typically, the animals exposed to such conditioned conflict show marked suppression of lever pressing in the presence of the tone, and one-hour experimental sessions, in which there were repeated presentations of the auditory punish-

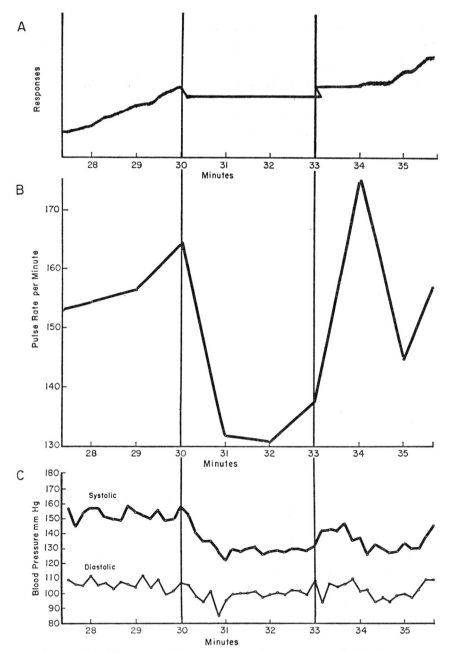

FIGURE 9 Changes in lever pressing, heart rate, and blood pressure during conditioned "anxiety" session after emotional conditioning.

ENDOCRINE SYSTEMS

ment signal, produced substantial 17-OH-CS elevations, as seen in Figure 10. In addition, determination of catecholamine levels in response to such conflict behavior revealed a hormone response pattern similar to that observed in the conditioned anxiety situation.

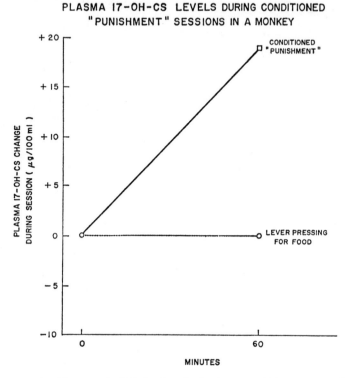

FIGURE 10 Plasma 17-OH-CS levels during conditioned "punishment" sessions. (From Mason, Brady, and Tolson[13])

Figure 11 shows the results obtained with two animals exposed to slightly different procedures using the conditioned conflict response. Catecholamine determinations were made of monkey A during a 10-minute conflict session, while such measurements with monkey B were made during a 10-minute period immediately preceding the conflict session, but after the animal had been unambiguously alerted as to what was to follow. In both instances, norepinephrine elevations occurred with no change in epinephrine levels.

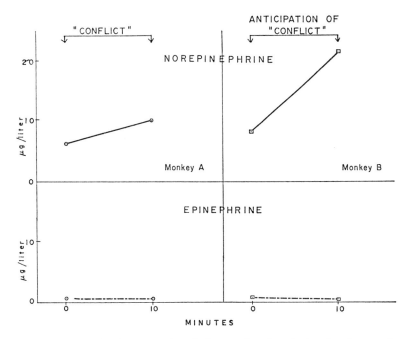

FIGURE 11 Plasma norepinephrine and epinephrine levels associated with conditioned "conflict" sessions. (From Mason, Brady, and Tolson[13])

In experiments involving more complex sequences of emotional behavior patterns with the monkey, however, it has been possible to observe differential changes in catecholamine levels under specified conditions. Figure 12, for example, summarizes the results obtained in an experiment during which the withdrawal of a blood sample 10 minutes prior to the start of a session produced marked elevations in both epinephrine and norepinephrine. In the course of previous conditioning trials, several different combinations of lever pressing for food alone, clicker-shock pairing alone, and concurrent presentation of lever pressing and clicker-shock pairing (the conditioned anxiety procedure) had been randomly programmed in such a way that the blood-withdrawal signal could not be predictably associated with any specific component of the sequence. Under these somewhat ambiguous circumstances, both epinephrine and norepinephrine levels rose significantly during the 10 minutes preceding the programed ses-

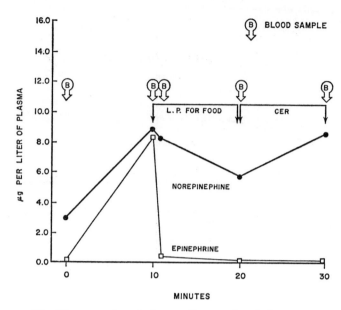

FIGURE 12 Plasma epinephrine and norepinephrine responses to ambiguous blood-withdrawal signal L.P., lever pressing; CER, conditioned emotional response. (From Mason, Mangan, Brady, Conrad, and Rioch[15])

sion, although epinephrine levels fell precipitously — from 8 micrograms per liter to 0.5 micrograms per liter — immediately after presentation of the first specific lever-pressing signal. A similar experiment, illustrated in Figure 13, used randomly programed 10-minute component segments of "time out," a shock-avoidance procedure to be described below, and the conditioned punishment or conflict situation described above. Extremely large epinephrine and norepinephrine responses were again observed during the initial 10-minute time-out component prior to the unpredictable onset of a specifically conditioned emotional behavior signal. Interestingly, both epinephrine and norepinephrine levels declined again after presentation of the first specific signal, even though in this case it required participation in a shock-avoidance task.

The extensive analysis of hormone response patterns associated with conditioned avoidance has been emphasized in previous reports on the psychophysiology of emotion.[2] Briefly, in the basic behavioral

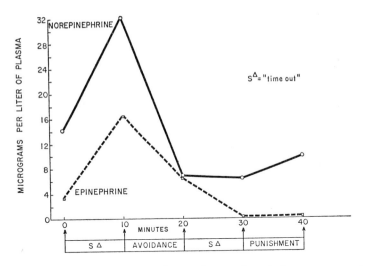

FIGURE 13 Plasma epinephrine and norepinephrine responses to randomly programmed components of a multiple schedule conditioning procedure. (From Mason, Mangan, Brady, Conrad, and Rioch[15])

procedure, shocks are programed to the feet every 20 seconds unless the animal presses the lever within that interval, postponing the shock another 20 seconds. This avoidance requirement generates a stable and durable lever-pressing performance in the monkey, and it has been shown to be associated consistently with twofold to fourfold rises in corticosteroid levels for virtually all animals during the two-hour experimental sessions, even in the absence of shock. It has also been possible to demonstrate quantitative relations between the rate of avoidance response in the monkey and the level of pituitary-adrenal cortical activity, independent of the shock frequency. Marked differences in the hormone response have been observed, however, when the avoidance procedure includes a discriminable exteroceptive warning signal presented 15 seconds after the previous response and five seconds prior to shock administration. Figure 14 compares the 17-OH-CS levels measured during "regular" and "warning signal" avoidance sessions with the monkey, and shows the consistently reduced corticosteroid response associated with programming such a warning signal. Conversely, superimposing so-called "free," or una-

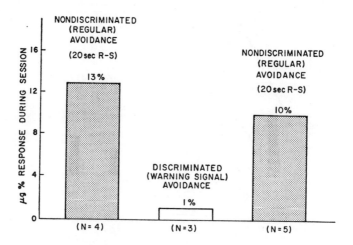

FIGURE 14 Plasma 17-OH-CS responses during nondiscriminated ("regular") and discriminated ("warning signal") avoidance sessions. N, numbers of animals in sample. (From Mason, Brady, and Tolson[13])

voidable, shocks upon a well-established avoidance baseline without a warning signal has been observed to produce elevations in 17-OH-CS. Figure 15, for example, shows that the presentation of such "free shocks" during two-hour avoidance sessions more than doubles the corticosteroid response as compared to the regular non-discriminated avoidance procedure.

Concurrent biochemical measurements of plasma corticosteroid and catecholamine levels have also been made in the course of several avoidance experiments with the monkey. The results, illustrated in Figure 16, confirm the previously described emotional behavior pattern of 17-OH-CS and norepinephrine elevations, with no significant alteration in epinephrine levels. However, two experimental manipulations with the avoidance procedure have been observed to produce significant variations in this hormone pattern. Figure 17 shows at least a modest epinephrine elevation with no change in norepinephrine; this accompanied presentation of the avoidance signal to a monkey well trained in avoidance after removal of the response lever from the restraining chair. Significantly, the effect occurred

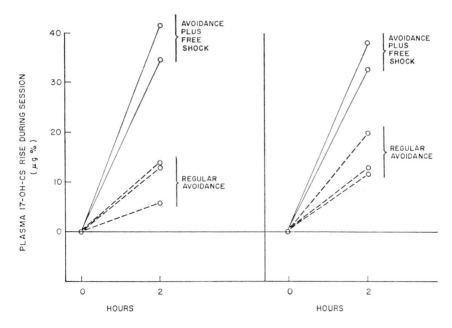

FIGURE 15 Plasma 17-OH-CS responses during "regular" nondiscriminated avoidance and during avoidance with "free shocks." (From Mason, Brady, and Tolson[13])

within one minute after the signal presentation, and could not be observed after 10 minutes of continued exposure.

The results obtained with the second series of experiments that produced such variations in catecholamine levels are illustrated in Figure 18, which shows the effects of free shock administration to a monkey at different stages in the course of avoidance training. The mild norepinephrine and epinephrine elevations shown in Figure 18A were obtained during an early conditioning session, during which more than 100 free shocks were given before the monkey had acquired the avoidance behavior. Section B shows the modest rise in norepinephrine levels with no change in epinephrine; this accompanied later experimental sessions that included performance of the well-learned avoidance response. Finally, Figure 18C shows the results of a series of experiments in which free shocks were programmed at the rate of one per minute (approximately the shock frequency occurring during

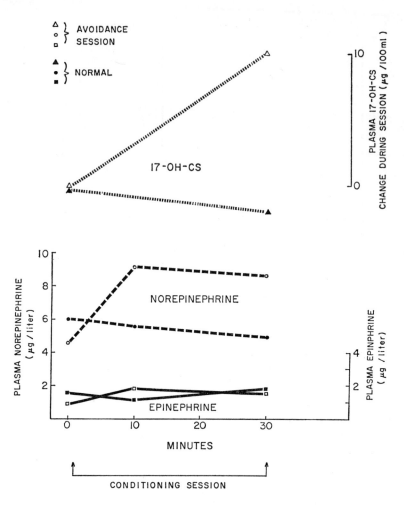

FIGURE 16 Mean plasma 17-OH-CS, norepinephrine, and epinephrine levels during conditioned avoidance sessions. (From Mason, Mangan, Brady, Conrad, and Rioch[15])

a typical avoidance session) with the same monkey. Significantly, dramatic elevations in both epinephrine and norepinephrine accompanied this procedural change, even though the animal received no more shock than during previous regular avoidance sessions.

Extended exposure to continuous 72-hour avoidance sessions has more recently provided the setting for an analysis of a broader spec-

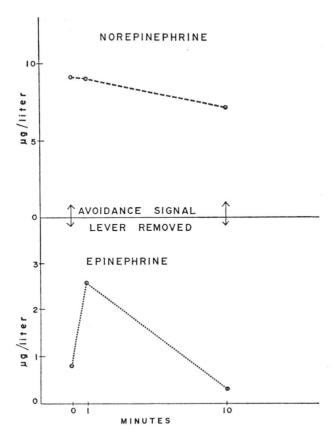

FIGURE 17 Plasma epinephrine and norepinephrine levels following removal of the lever during an avoidance session. (From Mason, Brady, and Tolson[13])

trum of hormonal changes in relationship to emotional behavior in the rhesus monkey.[11,14] The pattern of corticosteroid and pepsinogen levels was observed before, during, and after such an experiment. Although plasma 17-OH-CS levels showed the expected substantial elevation throughout the 72-hour session, plasma pepsinogen levels were consistently depressed below baseline values during this same period. The postavoidance recovery period, however, was characterized by a marked and prolonged elevation of pepsinogen levels. This continued for several days beyond the 48-hour postavoidance inter-

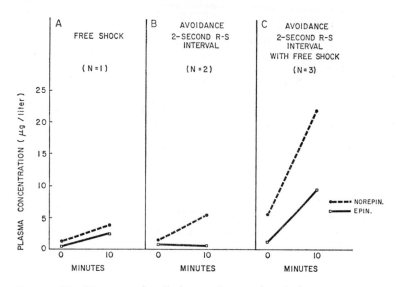

FIGURE 18 Plasma epinephrine and norepinephrine responses to "free shock" alone, "regular" nondiscriminated avoidance, and avoidance with "free shock." N, numbers of animals in sample. (From Mason, Brady, and Tolson[13])

val that was required for recovery of the preavoidance corticosteroid baseline. Patterns of thyroid, gonadal, and adrenal hormone secretion have most recently provided the focus for studies of the 72-hour avoidance situation. Observations have indicated that the endocrinological consequences of exposure to such conditions may be detectable for prolonged periods following their termination. In such experiments, 17-OH-CS and epinephrine levels have been found to rise to at least twice the baseline value during avoidance and to return to normal in from two to six days. Androsterone and estrone levels dropped to below half the baseline value during avoidance, but showed substantial rebound changes above baseline from three to six days after termination of the session. There was little change in the thyroid levels during the initial portions of the avoidance performance period, but a gradual elevation throughout the session produced a peak value early in the postavoidance recovery period, which in some instances did not return to baseline for a full three weeks after the end of the three-day session.

The results of these experiments establish firm relationships between a broad range of endocrine system activity and behavioral interactions in various aspects of emotion. The initial findings, which emphasize changes in absolute levels of selected hormones, can be viewed as reflecting relatively undifferentiated consequences of affective arousal states associated with effective emotional behavior patterns. The definite temporal course of steroid changes under such conditions, and the quantitative nature of the relationship between degree of behavioral involvement and level of corticosteroid response, have been well documented. In addition, the critical role of an organism's behavioral history in determining the nature and extent of endocrine participation in affective interactions has been convincingly demonstrated.

However, the most meaningful dimension for hormone analysis in relationship to more chronic affective interactions would appear to be the broader patterning or balance of secretory change in many interdependent endocrine systems, which in concert regulate metabolic events. Anatomically, points of articulation in the brain stem between nerve cells and the pituitary-adrenal system are well established, and similar central integrative mechanisms, involving the gonadal, thyroid, and posterior pituitary glands, are now in evidence. The extensive and prolonged participation of these fundamental endocrine systems in behavioral interactions suggests a relationship between such hormonal activity and the more durable aspects of feelings, which include generalized mood states, and affective dispositions. Indeed, the differentiation of such hormone response patterns in relation to the historical and situational aspects of behavioral events may well provide a first approximate step in the direction of identifying distinguishable intraorganismic affective consequences that are associated with both episodic and persistent emotional interactions.

The Conditions for Emotional Behavior

GEORGE MANDLER

One of the pleasures of discussing Dr. Brady's paper is that it gives me the opportunity to explore new directions and implications, without having to cavil about problems of experimental design, treatment of data, or interpretation. The data are unequivocal in demonstrating certain stable relations among emotional stimuli, emotional behavior, and psychoendocrine systems.

I will pursue four topics that his data and discussion have brought into focus for me. These all deal with the problem that is central to the psychologist's concern with emotional behavior: what are the conditions under which emotional behavior appears, what are the controlling stimuli, or, in a more traditional stance, what are the causes of emotional behavior?

1. What is the relation between psychological and physiological events and how can we best conceptualize different types of interactions between these two systems? In other words, what controls what?
2. Dr. Brady has made a distinction between feelings and emotional behavior. I would like to re-examine this distinction and to suggest an alternative interpretation, which does not make a radical division between these two sets of events.
3. I would like to review the body of Dr. Brady's data in terms of the control of the situation by the experimental subject. In particular, I will suggest an interpretation of "predictability" and "certainty" in terms of some of my own recent work.
4. I will pursue the previous topic by a brief discussion of some further data on the control of emotional behavior.

PSYCHOLOGICAL AND PHYSIOLOGICAL VARIABLES IN EMOTIONAL BEHAVIOR

Permit me to review a position I have taken elsewhere[1] on the relations between the physiology and the psychology of emotion. Spe-

---

GEORGE MANDLER University of California, San Diego

cifically, I want to make a distinction between physiological variables, per se, and *psychologically functional physiological variables*. In determining the controlling events for emotional behavior, the psychologist is primarily concerned with the latter set. Psychologically functional physiological variables are those which have a demonstrable controlling effect on behavior. In focusing on this class of variables, we must make a distinction between correlated and controlling variables. The former represent a vast collection from the work of both physiologists and psychologists. Correlated variables indicate that certain psychological and physiological events co-occur, but no statement can be made about any causal or controlling effects from one to the other. Controlling physiological variables, on the other hand, are those for which it "can be shown that the presence of one or more of them is either necessary or sufficient for variation in some behavior to occur."

This position can, of course, be reversed to apply to *physiologically functional psychological variables*, as some behavioral events presumably are the controlling events for physiological variation.

If we restrict ourselves to autonomic and endocrine events when discussing physiological variables, then the following four classes can be distinguished:

Psychologically functional physiological variables. These variables are of primary interest to the psychologist who is interested in physiological control of emotional behavior. I have reviewed the data on this particular problem[1] and still maintain that the list is rather meager. Many investigators have hunted this elusive set, but few have found it. What evidence there is suggests that the physiological variables controlling psychological events are gross, and that the minute and detailed variations we can measure in physiological response are not, in fact, the controlling stimuli for emotional behavior.

Psychologically nonfunctional physiological variables. Most psychophysiological research has only been able to specify these correlated physiological variables, many of which are important in their index functions, but any controlling effect has been shown for few of them. Most of the physiological measurements we have taken are, for the time being, in the psychologically nonfunctional category.

Physiologically functional psychological variables. To the extent that certain physiological events are controlled by behavioral events, this category is most fruitful for the current and future investigator. Relatively little work has been done in the area, but I think it offers a most promising approach, because previously slighted data on the relation between physiology and behavior may be given a second look.

Physiologically nonfunctional psychological variables. These behavioral functions have no effect on measurable physiological events.

Dr. Brady's experiments fall primarily in the area of correlated psychological and physiological variables. He does not present, nor does he claim to present, any evidence on the control of emotional behavior by the autonomic nervous system. He does, however, suggest that he is dealing with physiologically functional psychological variables when he says, for example, that steroid elevation is behaviorally induced or that the anxiety response eliminates the steroid response. Dr. Brady wants to talk only about covariants, but we must be careful not to draw the wrong conclusions, if only because this particular error has frequently been made in the literature. In the work Dr. Brady has discussed here, and in other work from his laboratory (cf. Sidman, et al.[9]), he presents some very nice evidence on the problem of physiologically relevant psychological variables. Sidman, et al., conclude that their data "clearly implicate[s] the avoidance behavior as an activator of the pituitary-adrenal cortical system." This kind of research is needed in order to specify some of the relations I have outlined above.

FEELINGS AND EMOTIONAL BEHAVIOR

Dr. Brady makes a distinction between "feelings" and "emotional behavior" that is not quite defensible. The term emotional behavior is to be reserved for that class of behavior that can be related directly to exteroceptive stimulus objects, while feelings are the consequences of interoceptive events. He suggests that the variety of feelings may be related to the large variety of different patterns of physiological events. This position suggests that feelings are psychological variables under the control of psychologically functional physiological variables, specifically physiological patterns. I disagree on two grounds:

first, there is no evidence at the present time that feelings are, in fact, under the control of different patterns of physiological events; second, there is a parsimonious alternative explanation which, in fact, makes feelings indistinguishable from "emotional behavior."

The evidence that physiological patterns control certain psychological events is meager, at best.[1] More important, however, one can argue, as Schachter[7] and I have argued, that the physiological events controlling feelings or any other emotional behavior are gross and undifferentiated. As a matter of fact, if one looks at just co-variational relations between physiological and psychological events, the best estimate of psychological effect is derived from a grouping of physiological variables.[3]

In short, I suggest that feelings are a joint function of gross undifferentiated, autonomic arousal and highly specific, highly differentiated environmental events. Given autonomic arousal, the organism seeks for an explanation of these events in his environment.[8] Thus, the controlling stimuli for highly differentiated feelings may be found in the differentiation of the cognitive or environmental events under which they take place.

This argument suggests that the human organism does not differentiate among patterns of autonomic events — although these patterns may have some behavioral effects we cannot at the present time specify — but that a gross autonomic arousal combined with some environmental events provides both feelings and emotional behavior. Feelings are just one sub-set of the same emotional behavior Dr. Brady discusses. I do not deny the importance of investigating and determining physiological patterns; all I suggest is that most of the behavioral correlates of emotion cannot at the present time be seen as a function of highly differentiated physiological patterns.

ANTICIPATION, CERTAINTY, AND CONTROL

Dr. Brady's relevant experimental data on anticipation, certainty, and control can be summarized as follows:

The CS (conditioned stimulus) for a noxious event alone (the horn) produces an NE (norepinephrine) but not an E (epinephrine) response. A signal for an event that the organism has previously encountered and responded to produces NE but not E.

When the organism is in a conflict or punishment situation, E stays level but NE is elevated. However, NE levels are higher during "anticipation," i.e., when the animal is unambiguously alerted to the required response.

Blood withdrawal produces both E and NE responses. Here the organism is presented with "uncertainty"; there is no single or simple response to be made to the blood-withdrawal CS, i.e., the situation is ambiguous.

In the avoidance situation, the warning signal reduces the level of 17-OH-CS. The warning signal is a signal for the emission of a previously practiced and well-established response.

When free shocks are administered, i.e., when there is no signal in the situation for a previously organized response, the 17-OH-CS response increases.

When the response lever is removed, the level of E increases. When the organism has no response available, i.e., when the situation is again ambiguous, the E level increases.

Both E and NE levels increase when the organism has no control over shock onset, i.e., when there is complete loss of certainty.

These data can be organized by a model of the organism's control over the situation. By control I mean simply that the organism has an available response that is relevant to the situation. If a warning signal occurs and the subject prepares for onset of shock, he is controlling that onset in the sense that shock becomes "subjectively" contingent on the response. When, however, events occur in the environment for which the organism has no behavior available, a perturbation of behavior and physiological evidence of distress or anxiety will result. The data indicate that with control over stimulus onset, in our terms, norepinephrine levels increase, while epinephrine levels increase when control is absent. In a previous summary of some of the data presented here, Mason, et al.,[5] stated that the NE response occurs when "conditions . . . are unambiguous and familiar," and that the E response involves an "element of uncertainty or unpredictability," a "necessity of an anticipatory set," or the "anticipation of coping activity." Rather than talk about "anticipation" and "necessities for sets," I would prefer to invoke the availability of relevant behavior. When an ongoing behavior sequence is interrupted and the organism does have some situationally relevant behavior available, anxiety will not appear, and when such behavior is not available, anxiety will appear.[2,4] It now seems that the interruption of behavior when the organism is "helpless"[6] effects an epinephrine response, while conditions in which relevant behavior is available to the organism produce a norepinephrine response.

DISTRESS IN THE ABSENCE OF AVAILABLE BEHAVIOR

As a final argument for the position that the unavailability of situationally relevant behavior produces distress (apparently an epinephrine response), I would like to review some recent data from our laboratory. We argued that behavioral disorganization or distress ought to appear when no noxious stimuli, per se, are used, but where only the availability of situationally relevant behavior is manipulated. In two animal studies,[2,4] we were able to produce extreme distress in the following situation.

Rats in a Y-maze learned to discriminate brightness. They were run hungry for 24 hours and were given a food pellet as a reward for the black-white discrimination. We then ran extinction trials under two different conditions. The first was the classical one in which the previous procedure was maintained throughout and the animals were 24-hours hungry when introduced to the maze. In the other condition, the animals were run satiated; they were put on an ad lib maintenance schedule and fed just prior to their introduction into the maze. Reward was absent in both groups.

Our argument was that the satiated animals are exposed to two conditions: the general stimulus complex of the maze, which causes them to continue the running response simply because it is the appropriate one elicited, and no alternative responses are available when they cannot complete the sequence of running through the maze and eating the pellet in the appropriate goal box. The deprived animals are not exposed to this dilemma, because the stimulus conditions of being hungry produce relevant and available food-searching behavior.

In fact, during extinction the satiated animals showed extremely excited behavior, which increased with successive trials. This excitement was apparently indicative of a high degree of distress. The behavior included jerks of the body and legs, and sometimes developed into convulsions that immobilized the animal for several seconds. Thus, these animals showed a high degree of distress when an organized sequence was interrupted and they had no situationally relevant behavior available. Let me say that in another experiment satiated animals that were given pellets in the goal box completed the run in all cases and ate the food pellets. Only the unrewarded satiated animals showed distress indexes.

In summary, I suggest that we abandon some of the vague notions of anticipation and ambiguity, and substitute for them the notion of availability of situationally relevant behavior. I submit that this is what we mean by anticipation. If an organism "anticipates" something, it has some available behavior in which it can engage when the proper signals appear. A situation is "ambiguous" when many different responses are available and no single set of behaviors is directly or dominantly relevant to the situation. The "unambiguous" situation is the one in which situationally relevant behavior is defined with little variance.

Dr. Brady's data show that the behavioral distinction of the availability of situationally relevant behavior is mirrored in the animals' physiological response. In terms of classes of variables discussed earlier, it appears that response availability is a physiologically relevant psychological variable.

Psychoendocrine Systems and Emotion: Biological Aspects

SEYMOUR S. KETY

Studies in operant behavior such as those which Dr. Brady has performed and described so well are impressive for their rigor and for the minimum of inferences in which they indulge. Such research has fulfilled an important need in psychobiology; our progress will be halting and unsure unless we are able to make small but sure steps from the firm foundation such studies provide. It is equally important, however, to look from time to time at the direction in which we are moving and at the goals which may perhaps motivate our steps. I should like to step back from the more molecular aspects and view the problem as a whole, freely admitting that many of the assumptions are speculative and the connections largely apparent.

It may be of value to begin the discussion of psychoendocrine relationships in emotional state by a consideration of biochemical methodology. Dr. Brady has used measurement of blood levels to great advantage; these are unexcelled for demonstrating rapid transients in the secretion of particular hormones that may occur over periods of a few minutes. Although they permit precise timing of changes in secretion rate, blood levels are difficult to interpret quantitatively or to compare with each other in terms of the actual secretion rate. Such levels for epinephrine reflect both the rate of secretion by the adrenal medulla and the rate of disappearance from the blood, which is largely by way of metabolism in the liver or excretion by the kidney. The half-time of epinephrine in the blood as opposed to that of the corticosteroids is quite short, so that the blood level is extremely sensitive to the rate of disappearance. That, in turn, depends on the blood flow to the kidney, liver, and splenic areas. The distribution of blood flow changes markedly in emotional states, so

---

SEYMOUR S. KETY Laboratory of Clinical Science, National Institute of Mental Health, National Institutes of Health, Bethesda, Maryland

this factor could become an overriding one in relation to the rate of secretion. Blood levels of norepinephrine reflect secretion of that hormone by the adrenal medulla as well as its release from various sympathetic nerve endings throughout the body, as well as factors that affect the distribution and the rate of destruction or excretion.

Because the catecholamines or their metabolites eventually end up almost entirely in the urine, measurement of urinary excretion of these substances is useful in terms of the higher concentrations which exist and the suitability of methods. Although the urinary excretion represents an integrated value of hormonal secretion over an hour or more, it is quite insensitive to, and of little value for, precise timing of transient changes. Until recently one faced another problem in the use of urinary excretion. Only five per cent of catecholamines released into the blood are excreted as such in the urine, and the exact percentage that escape metabolism is considerably influenced by the relative blood flow to the liver — the main seat of metabolism for the circulating catecholamines.

In recent years, however, the various metabolites of the catecholamines have been identified[1]: 3-methoxy-4-hydroxymandelic acid (vanillylmandelic acid, VMA), O-methylepinephrine (metanephrine), and O-methylnorepinephrine (normetanephrine). These three metabolites, plus the unchanged catecholamines, account for 80 to 90 per cent of the catecholamines that are secreted into the blood stream.[8] In addition, normetanephrine represents the first conversion product of norepinephrine when it is physiologically released at sympathetic nerve endings.[7] Thus, the normetanephrine:VMA ratio may tell something about general sympathetic activity. One recently discovered process, however, prevents the blood or urinary assay of norepinephrine and its products from accurately representing sympathetic nervous system activity. A considerable percentage of norepinephrine released at sympathetic nerve endings is taken back into these endings without passing into the blood stream or the urine, and that fraction of sympathetic activity is therefore lost to both these channels.

In spite of the problems enumerated above, however, blood or urine levels of catecholamines and their products are useful in terms of the timing or the rough quantification of changes in the rate at which these substances are secreted by the adrenal medulla or released by activation of the sympathetic nervous system.

The evidence Dr. Brady has presented, as well as that acquired in large numbers of other studies in animals and man, clearly indicate an increased secretion of epinephrine or norepinephrine in various types of what has usually been called stress. These include: parachute jumping, automobile racing, competitive sports, boxing, viewing emotion-laden movies, aggressive behavior, change from salaried to piecework conditions.[2,3,9,10] The data furthermore sustain a generalization that supports an important thesis of Schachter[11]: the release of catecholamines is related to the intensity rather than to the quality of affect, while the nature of the affect depends to a considerable extent upon cognitive factors and the past experience or present situation of the individual subject. Thus, Levi found an increase in urinary catecholamine excretion in subjects watching certain motion pictures, and the increase was quite the same whether they were viewing a hilarious comedy or a horror movie.

One of the most interesting findings to which Dr. Brady alluded is the differential release of epinephrine and norepinephrine under different circumstances. Physiological mechanisms exist to account for this, and differences in the action of the two catecholamines contribute to an understanding of the remarkable adaptation this exemplifies. Folkow and Euler[4] showed a selective activation of norepinephrine- and epinephrine-producing cells in the adrenal upon stimulation of different areas of the hypothalamus. These substances also differ in their peripheral effects. Epinephrine mobilizes glucose into the blood stream, dilates the arterioles of the heart, brain, and skeletal muscle, speeds the heart and increases its output. The chief effect of norepinephrine is to constrict arterioles generally and raise the blood pressure. It has considerably less effect on blood glucose and upon the action of the heart than does epinephrine. Thus, epinephrine would be of great use to the organism and the muscular exercise required in flight or fight, whereas norepinephrine would be of advantage in protecting the animal from the acute effects of hemorrhage.

In addition to Dr. Brady's interesting observations of animals, there are the findings of some of his and my colleagues who studied the blood levels and urinary excretion of these two catecholamines in a number of clinical situations. Immediately upon admission to a hospital, both catecholamines were elevated,[12] but, surprisingly enough, on the morning of a surgical operation only norepinephrine was elevated.

Both types of observation in conjunction with the different physiological effects of these amines suggest the generalization[5] that epinephrine is secreted primarily in situations of uncertainty, in which flight or fight may be the appropriate response, and this agent would have adaptive functions. Norepinephrine appears to be secreted in those situations in which the outcome is inevitable or unavoidable and muscular activity would be inappropriate or useless.

The adaptive value of the steroid hormones released from the adrenal cortex under stress is somewhat less clear. It is likely that their anti-inflammatory role, their ability to stimulate wound healing, and their other reparative effects would have significant survival value. In addition, the corticosteroids have some interesting relationships with the catecholamines. The release of both types of humoral agent is under the control of hypothalamic centers. That of the corticosteroids is mediated by adrenocorticotropic hormone (ACTH) and that of the catecholamines by the sympathetic nervous system. There is evidence indicating that epinephrine, acting on the hypothalamus, causes the release of ACTH, and only recently Wurtman and Axelrod[13] in our laboratory have found evidence for an interesting feedback in the reverse direction — an effect of circulating adrenal steroids on the enzyme responsible for the synthesis of epinephrine in the medulla (phenylethanolamine-N-methyl transferase). ACTH, through its release of corticosteroids and their stimulation of the synthesis of this enzyme, causes an increase in epinephrine production in the adrenal medulla. The portal circulation from the adrenal cortex to the medulla that has been described, and the close proximity of these two seemingly different endocrine glands suggest an important physiological control of one upon the other.

In contrast to the reasonably well-defined role catecholamines in the periphery play in emotional behavior and response to stress, their possible role in the brain in affective or emotional states is largely speculative, although evidence for their involvement is increasingly convincing. Norepinephrine is one of several amines that are found in relatively high concentrations in the brain, localized especially in the hypothalamus and other areas of the limbic system; this, in fact, has suggested their involvement in affect and emotion. There is little disagreement that drugs which deplete the brain of amines are depressant agents, and that these effects are in some way related. Drugs that inhibit monoamine oxidase, the enzyme responsible for the destruction of these amines, will permit their accumulation in the brain and at the same time act as antidepressant agents in man or as excitants to animals, when given in large dosages. All of the drugs that have significant effects on mood seem to have one or another effect upon brain norepinephrine — they either deplete it and produce depression, or favor its release or accumulation at appropriate receptor sites in the brain in association with antidepression, euphoria, or hyperactivity.[6]

It does not seem entirely premature to suggest that the evidence strongly indicates that norepinephrine in the brain plays an important role in mediating alertness and wakefulness, pleasure and euphoria, anger and fear. Although the evidence implicating norepinephrine in some of these emotional states is good, that does not preclude the operation of other amines, such as serotonin or dopamine, whose actions have not been as well examined. Furthermore, it is extremely unlikely that a specific amine will ever be found to be entirely responsible for a specific emotional state. It is more likely that the important determining factors will be an interaction among certain amines at particular sites and, most important, in association with particular cognitive factors, which, taken all together, may define a particular emotion. Thus, Schachter's demonstration of the importance of cognitive factors in the emotional state aroused by epinephrine in the periphery seems to have even greater cogency to the chemical substrate for emotion centrally. It appears quite likely that, at best, the interrelationship of the biogenic amines with particular sites may determine the intensity of emotional state, but its quality will probably always be dependent to a considerable extent upon the special idiosyncratic significance of the setting and the past experience of the particular individual.

Involvement of the Hormone Serotonin in Emotion and Mind

D. WAYNE WOOLLEY

Dr. Brady has told us how emotions of fear and feelings of anxiety can cause changes in the amounts of certain hormones in the blood. I would like to continue the discussion by showing how changes in a specific hormone (viz., serotonin) can bring about changes in emotions and even of intellect.

Serotonin is one of the few hormones for which there is substantial evidence to indicate that it participates directly in the formation of emotion and the functioning of intellect. These participations are often manifested by changes in behavior. It is not only that serotonin occurs in the brain and can be shown to exert actions on various neurophysiological functions. These facts permit us to distinguish between a mere pharmacological curiosity and something which may have real physiological meaning, but in itself, the occurrence of a substance in brain does not necessarily show us that it is related to intellect. The establishment of this relationship requires more than can be accomplished by electrophysiological measurements.

In this paper I want to mention very briefly five kinds of evidence

---

D. WAYNE WOOLLEY   The Rockefeller University, New York

which link serotonin and its functioning in the brain to the generation and control of emotions and other aspects of mind. Each of these five points of evidence can only be mentioned, because time does not permit an adequate presentation of any of them. For those who wish to pursue the matter in more detail, the evidence is presented in a book by D. W. Woolley, entitled "The Biochemical Bases of Psychoses."[4]

The original evidence which suggested that serotonin was concerned with mental processes was the discovery in 1954[6,7] that mental aberrations very similar to those found in schizophrenia could be called forth in normal human beings by the ingestion of any one of a series of structural analogs of this hormone. These analogs were known to owe much of their effects on other tissues to their ability to interfere with the actions of serotonin, and consequently it seemed possible that their effects on intellect might also arise from an interference with the actions of serotonin in the brain.

By suitable pharmacological and biochemical experiments it can be shown that these analogs of serotonin can exert two kinds of actions against the hormone. Some of them can prevent the actions of serotonin, presumably by specific blockage of the serotonin receptors in the manner well established for the action of other antimetabolites. Other analogs are sufficiently akin to serotonin in chemical structure to take the place of the hormone, and exert effects on tissues very similar to those caused by the hormone itself.

The serotonin analogs in which the hormone-like action is prominent usually cause symptoms of excitement and agitation in normal persons. The analogs in which the antagonism to serotonin is prominent tend to cause mental depression. It is with the analogs with marked serotonin-like effects that one can call forth in normal persons visual or auditory hallucinations, sometimes accompanied by sensations of pleasure and often by mental aberrations.

Let us look briefly at a few of these psychotomimetic serotonin analogs. Figure 1 shows a few of the synthetic compounds with which this phenomenon was first encountered. Medmain has never been tested on human beings because its ability to cause excitement in laboratory animals was so marked that a human test was inadvisable.

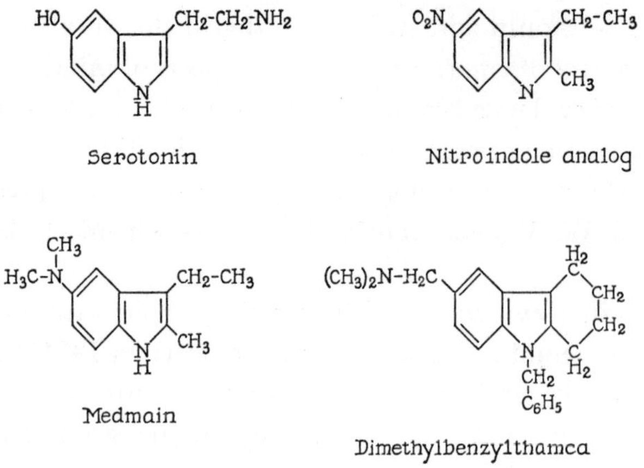

FIGURE 1 Serotonin and three of its analogs that show effects on mental activity.

The same thing can be said for the benzyldimethylthamca, which proved to be so violent in its action on dogs that the animals were obviously deranged for several days following its administration. Before these compounds had been made, however, the nitro analog shown in the figure had been taken by human beings, and it had been observed to cause a temporary, but severe, mental depression.

Let us turn now from these synthetic analogs to some which occur naturally. Many of these have been examined, but two will serve to illustrate the point under discussion. These are psilocybin and LSD. The chemical structure and resemblance of these drugs to serotonin is illustrated in Figure 2. These are serotonin analogs which cause various kinds of excitement and mental change in normal persons. LSD has been much used to induce visual hallucinations and temporary changes in personality. It is the most active of the known psychotomimetic drugs, in that less than 0.1 milligram is sufficient to call forth an effect in a man. Psilocybin has been used for centuries in Yucatan to induce keener insight and a trancelike state.

Both psilocybin and LSD have been shown to act like serotonin on a variety of tissues, including certain parts of the brain. In addition to this serotonin-like effect, they can also act as specific inhibitors against serotonin in certain kinds of tissue.

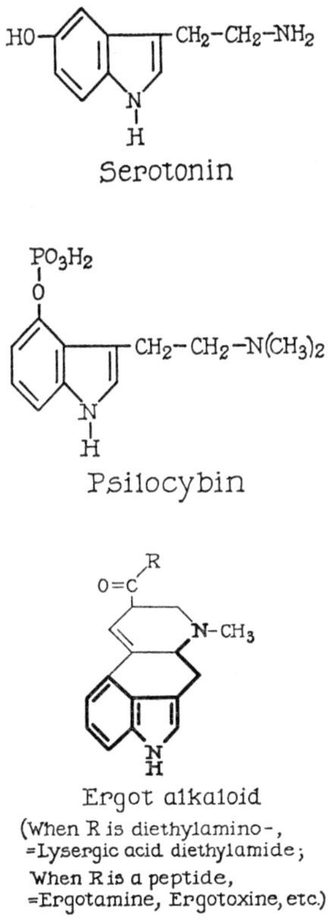

HO–[indole]–CH$_2$–CH$_2$–NH$_2$

Serotonin

PO$_3$H$_2$–O–[indole]–CH$_2$–CH$_2$–N(CH$_3$)$_2$

Psilocybin

O=C(R)–...–N–CH$_3$ [ergoline ring system]

Ergot alkaloid
(When R is diethylamino-,
=Lysergic acid diethylamide;
When R is a peptide,
=Ergotamine, Ergotoxine, etc.)

FIGURE 2 Chemical structures of psilocybin and LSD show resemblance to that of serotonin.

To sum up this first type of evidence, it is plain that marked mental changes of a variety of kinds can be caused by administration of serotonin analogs. These analogs act like serotonin, and some of them act against it as specific antagonists. Those which act like the hormone can cause excitement, while those which act more as antagonists may cause depression. These are abbreviated summations that skirt some of the complexities of the real situation.

The second kind of evidence that linked serotonin to mental processes was the discovery that some of the tranquilizing drugs were able to cause deficiencies of serotonin in the brain and other tissues. This was first shown for reserpine by Brodie and his collaborators.[2]

The structure of reserpine is given in Figure 3, where it is possible to see that it is an analog of the hormone. Chemical analysis of tissues of animals treated with reserpine show that it has caused a large reduction in the content of serotonin. The drug seems to have the ability to prevent the uptake of serotonin into the storage vesicles, and may also have a disruptive action on these vesicles. Larger doses

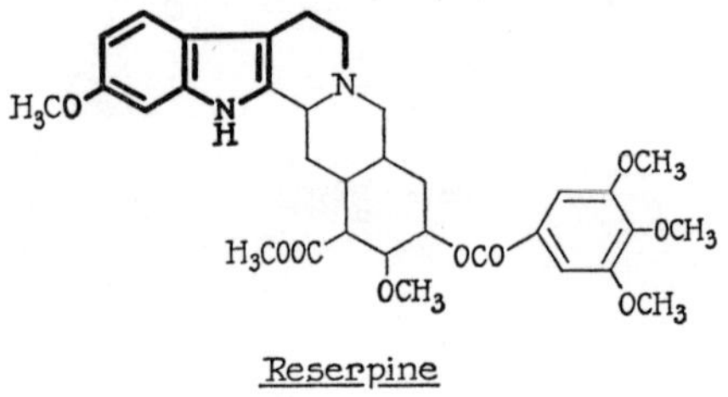

Reserpine

FIGURE 3   Structure of reserpine demonstrates that it is an analog of serotonin.

can interfere at other receptors for the action of serotonin. Its ability to cause tranquilization has been directly related to its ability to reduce the serotonin content of the brain.

Chlorpromazine, another widely used tranquilizing drug shown in Figure 4 does not deplete tissues of serotonin. However, it combines

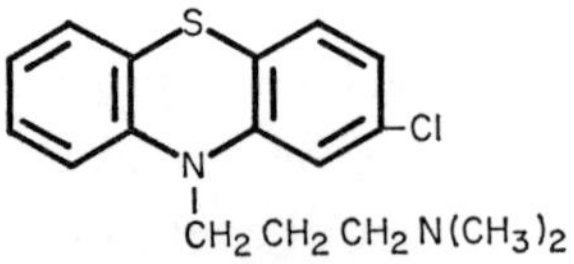

FIGURE 4   Chlorpromazine, a widely used tranquilizing drug, causes functional lack of serotonin.

with the serotonin receptors in susceptible tissues, and renders these receptors incapable of responding to serotonin. The net result is thus quite similar to a loss of the hormone. Chlorpromazine may be said to cause a *functional* lack of serotonin in susceptible tissues.

In addition to the actual or functional deficiences of serotonin caused by these two tranquilizing agents, they bring about similar changes with respect to the catecholamines. Thus, reserpine depletes tissues of norepinephrine as well as of serotonin. Chlorpromazine blocks the receptors for epinephrine and for norepinephrine as well as for serotonin. This has introduced discussion on which hormone is responsible for the mental effects—a discussion it is not possible to deal with in this short paper. There is, however, full agreement that the effects relative to serotonin represent the chemical groundwork for some of the behavioral and emotional effects of the drugs.

The third kind of evidence that relates serotonin to emotion and subsequent behavior is that the artificial increase of serotonin in the brain brings about psychic changes, which can perhaps be summed up with the words "elation" or "euphoria." The increase can be accomplished either by peripheral administration of the biochemical precursor of the hormone (viz., 5-hydroxytryptophan), or by use of a drug to prevent the destruction of the hormone. Such a drug is one of the inhibitors of monoamine oxidase.

Figure 5 illustrates some of the biochemical reactions by means of which serotonin in the brain is formed, and, after its effect has been exerted, is destroyed. The effective concentration of the hormone can be increased either by administration of the precursor (viz., 5-hy-

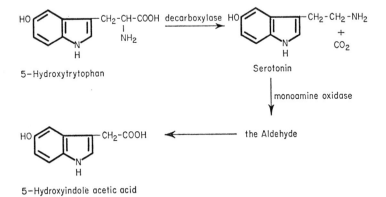

FIGURE 5 Major pathway of formation and destruction of serotonin in mammals.

droxytryptophan), which causes more to be formed, or by inhibition of the destructive enzymes with an inhibitor of monoamine oxidase.

Although the induction of euphoria by inhibitors of monoamine oxidase was an accidental discovery, the use of these drugs to relieve many patients suffering from simple mental depressions stems from the work of Kline and of Zeller.[3] More recently, after the metabolism of serotonin was elucidated, and its importance for mental processes became more firmly established, the use of 5-hydroxytryptophan for the relief of mental depressions has been studied. Enthusiastic clinical reports on the effectiveness of this treatment have recently appeared.

The fourth kind of evidence to suggest that serotonin plays an important role in the development of the intellect has come from a study of the disease known as phenylketonuria. This inborn error of metabolism is found in those human beings who, for genetic reasons, do not possess the enzyme which converts dietary phenylalanine to tyrosine. Such persons almost always show mental retardation early in life. Once the damage has taken place it is irreparable.

The mental defect is, however, preventable if, early in infancy, the patient is fed by a diet low in phenylalanine. That the mental failure can thus be prevented merely by maintenance of the body levels of phenylalanine somewhere near the normal concentration shows clearly that brain damage is the result of excess tissue levels of this amino acid during early infancy.

Why should excess phenylalanine produce permanent damage to the developing mind? There is now clear evidence to show that in phenylketonuria the tissues are deficient in serotonin and in catecholamines.[1] This is believed to be caused by the inhibition of the decarboxylating enzyme that synthesizes these hormones. The inhibition is caused by metabolic products derived from phenylalanine, such as phenylpyruvic acid. The inhibition by these substances can be demonstrated both in vitro and in vivo, but there is still much debate as to whether the hormonal deficiency is entirely the result of this inhibition, or whether there are other contributing enzymic suppressions.

Phenylketonuria can be produced in experimental animals by feeding them large quantities of phenylalanine. When the disease was so produced and maintained in infant mice, it was possible by a suitable test of maze-learning ability to show that they were less able to learn than were normal mice. This defect in performance could be prevented by administration of a suitable derivative of serotonin.[8] Data to illustrate this point are shown in Figure 6. The prevention

Treatment	No. of mice	Av. score
Controls	105	7.5
DL-Phenylalanine + L-tyrosine (P + T)	91	6.3
P + T + melatonin	103	7.2
P + T + HTP	17	7.2

FIGURE 6 Prevention of the learning deficit in maze test of phenylketonuric mice with serotonin derivatives. All compounds were given continuously from birth: melatonin, 10γ, DL-5-hydroxytryptophan, 10 to 100γ, per gm mouse per day.

of the cerebral serotonin deficiency by administration of melatonin or 5-hydroxytryptophan largely prevented the defect in learning ability.

The fifth kind of evidence comes from the discovery of a serotonin synergist in the tissues of schizophrenic persons. Schizophrenia has been said to be a defect in the connections between thinking and emotions. If this is true, a discovery of a biochemical cause of the disease would give strong presumptive evidence that the chemical substance involved did, in fact, have much to do with emotion, mind, and consequent behavior.

No chemically demonstrable defect in the amount of serotonin in the tissues of schizophrenic patients has yet been demonstrated, although the question of whether it occurs has not been adequately explored. However, very recently a substance that sensitizes normal tissues to the effects of this hormone has been found in the tissues of many schizophrenic patients. This serotonin synergist seems to be present in elevated amounts in schizophrenic tissues, although small quantities of it are found in normal tissues. These elevated amounts, if they should prove to be characteristic of schizophrenia, would allow one to understand how, in this disease, the normal concentration of the hormone could be acting like an excess of it, thereby causing aberrations of emotion and behavior. Chemical work on this substance suggests that the synergist is a ganglioside. When one reflects on the recent demonstration that the serotonin receptor of brain also is a specific ganglioside,[5] the idea arises that possibly the emotional and mental aberrations of schizophrenia may be in part the result of a defect in the receptor mechanism for the hormone serotonin.

To summarize, five kinds of evidence combine to suggest that some aspects of emotion and of cognitive function seem to be under the influence of serotonin. There is more evidence to link this hormone to these matters than for any other hormone, with the possible exception of thyroxine. These five points of evidence are: 1) the psychotomimetic analogs of serotonin; 2) the tranquilizing drugs related to serotonin; 3) psychic effects of increases in cerebral serotonin; 4) serotonin deficiency in phenylketonuria; 5) a serotonin synergist in schizophrenic patients.

Cognitive Effects on Bodily Functioning: Studies of Obesity and Eating

STANLEY SCHACHTER

DISCUSSION

Some Psychophysiological Considerations of the Relationship Between the Autonomic Nervous System and Behavior
MARVIN STEIN   145

Inside Every Fat Man
NORMAN A. SCOTCH   155

Although we rarely bother to make the matter explicit, the assumption of an identity between a physiological state and a psychological or behavioral event is implicit in much contemporary work in such areas as psychopharmacology, psychophysiology, or any domain concerned with the relationship of bodily state to emotion or to behavior. Simply put, much of this work seems to proceed on the assumption that there is a simple, one-to-one relationship between a biochemical change or a physiological process and a specific behavior. It is as if we assumed that physiological state is an "unconditionally sufficient condition" to account for a psychological event.

Such an assumption has, of course, been enormously fruitful in many areas of purely biological and medical research. Spirochetes cause syphilis. Kill the spirochete and cure syphilis. An iodine-deficient diet leads to colloid goiters; repair the deficiency, repair the goiter. As one moves from the world of purely medical and physio-

---

STANLEY SCHACHTER Department of Social Psychology, Columbia University

logical research, however, the assumption of such an "identity" seems to become more and more troublesome. It is this assumption, for example, which is the crux of the James-Cannon difficulties. James' view of emotion rested squarely on the assumption of an identity between physiological and emotional state, and Cannon's brilliant critique of the James-Lange theory was, in essence, an attack on this assumption. It is this implicit assumption which is, I suspect, responsible for the impression of utter confusion in an area such as psychopharmacology, where it sometimes seems the rule rather than the exception to find a single drug proved in a variety of studies to have blatantly opposite behavioral effects. LSD, for example, has been proved to be a hallucinogenic and a non-hallucinogenic, to be a euphoriant, a depressant, and to have no effects on mood at all. This nightmarish pattern of conflicting and nonreplicable results is familiar to anyone who has delved into the literature on behavioral or "emotional" effects of many of the so-called psychotropic drugs. The pattern, however, is not limited only to the exotic drugs; even as familiar an agent as adrenalin has a similarly depressing history. Many years ago the endocrinologist Marañon[8] injected several hundred of his patients with adrenalin and then asked them to introspect. Some of his subjects simply described their physical symptoms and reported no emotional effects at all; others described their feelings in a fashion that Marañon labeled the "cold," or "as if," emotions; that is, they made statements such as "I feel *as if* I were afraid," or "*as if* I were awaiting a great happiness." Still other subjects described themselves as feeling genuine emotions. Of those who noted any emotional effects at all, some described themselves as feeling anxious, some as angry, some as euphoric. In short, adrenalin, producing almost identical and typical physiological effects in most of these subjects, produced a wide diversity of self-reports of feeling states. This situation is, I suspect, inevitable and will remain puzzling and discouraging as long as we persist in the assumption of an identity between the physiological and the psychological effects of a drug. If we do, my guess is that we will be just about as successful at deriving predictions about complex behavior from a knowledge of biochemical and physiological conditions as we would be at predicting the destination of a moving automobile from an exquisite knowledge of the workings of the internal combustion engine and of petroleum chemistry.

If we are eventually to make sense of this area, I believe we will be forced to adopt a set of concepts with which most physiologically inclined scientists feel somewhat uncomfortable and ill-at-ease, for they are concepts which are, at present, difficult to physiologize about or to reify. We will be forced to examine a subject's perception of his bodily state and his interpretation of it in terms of his immediate situation and his past experience. We will be forced to deal with concepts about perception, about cognition, about learning, and about the social situation.

In order to avoid any misunderstanding, let me make completely explicit that I am most certainly not suggesting that such notions as perception and cognition do not have physiological correlates. I am suggesting that at present we know virtually nothing about these physiological correlates, but that we can and must use nonphysiologically anchored concepts if we are to make headway in understanding the relations of complex behavioral patterns to physiological and biochemical processes.

To move from generalities, let us consider the effects of adrenalin or epinephrine. We know that an injection of $\frac{1}{2}$ cc. of a 1:1000 solution of epinephrine causes an increase in heart rate, a marked increase in systolic blood pressure, a redistribution of blood with a cutaneous decrease, and a muscle and cerebral blood-flow increase. Blood sugar and lactic acid concentrations increase and respiration rate increases slightly. As far as the human subject is concerned, the major subjective symptoms are palpitation, slight tremor, and sometimes a feeling of flushing and accelerated breathing.

These are some of the measured physiological effects of an injection of epinephrine. In and of themselves are such bodily changes pleasant or unpleasant? Given these symptoms, should the subject describe himself as angry, or as anxious, or as manic or euphoric, or simply as sick? From the results of the Marañon study, any of these self-descriptions are possible. How can we make coherent sense of such findings?

Several years ago, bemused by such results, my colleagues and I undertook a program of research on the interaction of physiological and cognitive determinants of emotional state. This program was based on speculation about what was, at that time, a hypothetical event. Imagine a subject whom one somehow managed to inject covertly with adrenalin, or to feed a sympathomimetic agent, such as ephedrine. Such a subject would become aware of palpitations, tremor, etc., and at the same time be utterly unaware of why he felt this way. What would be the consequences of such a state?

In other contexts,[12] I have suggested that precisely this condition would lead to the arousal of evaluative needs; that is, pressures would operate on such an individual to understand and evaluate his bodily feelings. His bodily state roughly resembles the condition in which it has been at times of emotional excitement. How would he label his present feelings? I would suggest that such an individual would label his bodily feelings in terms of the situation in which he finds himself. Should he at the time be watching a horror film, he would probably decide that he was badly frightened. Should he be with a beautiful woman, he might decide that he was wildly in love or sexually excited. Should he be in an argument, he might explode in fury and hatred. Or, should the situation be completely inappropriate, he could decide that he was excited or upset by something that had recently happened. In any case, it is my basic assumption that the labels one attaches to a bodily state, how one describes his feelings, are a joint function of such cognitive factors and of a state of physiological arousal.

This line of thought, then, leads to the following propositions:

Given a state of physiological arousal for which an individual has no immediate explanation, he will "label" this state and describe his feelings in terms of the cognitions available to him. To the extent that cognitive factors are potent determiners of emotional states, it could be anticipated that precisely the same state of physiological arousal could be called "joy" or "fury" or any of a great diversity of emotional labels, depending on the cognitive aspects of the situation.

Given a state of physiological arousal for which an individual has a completely appropriate explanation (e.g., "I feel this way because I have just received an injection of adrenalin"), no evaluative needs will arise and the individual is unlikely to label his feelings in terms of the alternative cognitions available.

Given the same cognitive circumstances, the individual will react emotionally or describe his feelings as emotions only to the extent that he experiences a state of physiological arousal.

The experimental test of these propositions requires, first, the experimental manipulation of a state of physiological arousal or sympathetic activation; second, the manipulation of the extent to which the subject has an appropriate or proper explanation of his bodily state; and third, the creation of situations from which explanatory cognitions may be derived.

In order to satisfy these requirements, Jerome Singer and I[17] constructed an experiment that was cast in the framework of a study of the effects of vitamin supplements on vision. As soon as a subject arrived, he was told: "In this experiment we would like to make various tests of your vision. We are particularly interested in how a vitamin compound called Suproxin affects the visual skills. If you agree to take part in the experiment we would like to give you an injection of Suproxin."

If a subject agreed (and all but one of the 185 subjects did), he received an injection of one of two forms of Suproxin—placebo or epinephrine. We have, then, two groups of subjects—placebo subjects on whom the injection can have no possible effects, and epinephrine subjects who, within a few minutes after injection, will become aware of the full battery of sympathomimetic symptoms.

In order to manipulate the extent to which subjects had a proper explanation of their bodily state, those who received epinephrine received one of two types of instructions.

Informed subjects. Before receiving the injections, such subjects were told, "I should also tell you that some of our subjects have experienced side effects from the Suproxin. These side effects will only last for 15 or 20 minutes. Probably your hands will start to shake, your heart will start to pound, and your face may get warm and flushed."

These subjects, then, are told precisely what they will feel and why they will feel it. For such subjects, the evaluative needs are low. They have an exact explanation for their bodily feelings, and cognitive or situational factors should have no effects on how the subject labels his feelings.

Uninformed subjects. Such subjects are told that the injection will have no side effects at all. These subjects, then, will experience a state of sympathetic arousal, but the experimenter has given them no explanation for why they feel as they do. Evaluative needs then should be high, and cognitive-situational factors should have maximal effect on the way such a subject labels his bodily state.*

Finally, in order to expose subjects to situations from which they might derive explanatory cognitions relevant to their bodily state, they were placed in one of two situations immediately after injection:

Euphoria. A subject was placed alone in a room with a stooge who had been introduced as a fellow subject and who, following a completely standardized routine, acted in a euphoric-manic fashion, doing such things as flying paper airplanes, hula-hooping, and the like, all the while keeping up a standard patter and occasionally attempting to induce the subject to join in.

Anger. A subject was asked to fill out a long, infuriatingly personal questionnaire that asked such questions as:
"With how many men (other than your father) has your mother had extramarital relationships?"
4 and under_____: 5–9_____: 10 and over_____."

Filling in the questionnaire alongside the subject was a stooge, again presumably a fellow subject, who openly grew more and more irritated at the questionnaire and who finally ripped the thing up in a rage, slammed it to the floor while biting out, "I'm not wasting any more time; I'm getting my books and leaving," and stamped out of the room.

In both situations, an observer, watching through a one-way mir-

---

* For purposes of brevity, the description of this experiment does not include details of all the conditions in this study. The chief omission is a description of a control condition introduced to evaluate alternative interpretations of the data. The interested reader is referred to the original paper by Schachter and Singer (1962).[17]

ror, systematically recorded the behavior of the subject in order to provide indexes of the extent to which the subject joined in the stooge's mood. Once these rigged situations had run their course, the experimenter returned and, with a plausible pretext, asked the subject to fill out a series of standardized scales to measure the intensity of anger or euphoria.

We have, then, a set of experimental conditions in which we are simultaneously manipulating the degree of sympathetic arousal and the extent to which subjects understand why they feel as they do, and measuring the impact of these variations on the extent to which the subject catches the mood of a situation rigged to induce euphoria in one set of conditions and to induce anger in another. From the line of thought that generated this study, it should be anticipated that subjects injected with epinephrine and told that there would be no side effects should catch the mood of the rigged situation to a greater extent than subjects who had been injected with a placebo or those who had been injected with epinephrine and given a completely appropriate explanation of what they would feel and why.

Examining first the results of the euphoria conditions, we find that this is exactly the case. The uninformed epinephrine subjects — those who had been told that there would be no side effects — tend to catch the stooge's mood with alacrity; they join the stooge's whirl of activity and invent new manic activities of their own. In marked contrast, the informed epinephrine subjects and the placebo subjects who give no indication of autonomic arousal tend simply to sit and stare at the stooge in mild disbelief. The relevant data are reported in detail elsewhere. For present purposes it should suffice to note that these differences between conditions are large and statistically significant on both observational and self-report measures of mood.

In the anger conditions, the pattern of results is precisely the same. Uninformed epinephrine subjects grow openly annoyed and irritated, while placebo and informed epinephrine subjects maintain their equanimity. The evidence is good, then, in support of our basic propositions. Given a state of physiological arousal for which a subject has no easy explanation, he proves readily manipulable into the disparate states of euphoria and anger. Given an identical physiological state for which the subject has an appropriate explanation, his mood is almost untouched by the rigged situation.

Such results are not limited to the states of anger and euphoria. In still other experiments in which similar techniques and comparisons were employed, we have been readily able to manipulate uninformed epinephrine subjects into amusement, as measured by laughter at a slapstick movie,[18] and into fearful or anxious states.[16]

In sum, precisely the same physiological state — an epinephrine-induced state of sympathetic arousal — can be manifested as anger, euphoria, amusement, fear, or, as in the informed subjects, as no mood or emotion at all. Such results are virtually incomprehensible if we persist in the assumption of an identity between physiological and psychological states, but they fall neatly into place if we specify the fashion in which cognitive and physiological factors interact. With the addition of cognitive propositions, we are able to specify and manipulate the conditions under which an injection of epinephrine will or will not lead to an emotional state and to predict what emotion will result.

These demonstrations of the plasticity of interpretation of bodily state have depended upon the experimental trick of manipulating physiological and cognitive factors simultaneously and independently. In nature, of course, cognitive or situational factors trigger physiological processes, and the triggering stimulus usually imposes the label we attach to our feelings. We see the threatening object; this perception-cognition initiates a state of sympathetic arousal and the joint cognitive-physiological experience is labeled "fear."

Several considerations suggest that the line of reasoning guiding these experimental studies of emotion may be extended to such naturally occurring states, and that the intensity of such states may be as modifiable as are experimentally induced states of arousal. As an example of this possibility, consider pain. Broadly, we can conceive of the intensity of experienced pain and of one's willingness to tolerate pain as a function of the intensity of stimulation of the pain receptors, of the autonomic correlates of such stimulation, and of a host of cognitive and situational factors. To the extent that we can convince a subject undergoing electric shock that his shock-produced symptoms and arousal state are caused not by shock, but by some outside agent such as a drug, he should, following the above considerations, experience less pain and be willing to tolerate more shock. Such an individual would, of course, regard his arousal as a drug-produced state, rather than as an indicator of pain or fear.

In an experiment designed to evaluate these expectations, Nisbett and Schachter[9] tested subjects' tolerance for a graded series of electric shocks. There were, in essence, two conditions. In one, the subjects took a placebo pill and were told that the side effects of the pill would be palpitations, hand tremor, breathing rate changes, and a sinking feeling in the pit of the stomach — all symptoms which pretests had shown actually accompanied anticipation and receipt of shock. In a second condition, the subjects also received a pill, but the side effects described (e.g. itching skin, numb feet, etc.) had nothing in common with the physiological symptoms accompanying shock. Ten minutes after taking the pill, both groups of subjects were given a series of brief shocks that systematically increased in intensity. They were told to tell the experimenter when the shock was too painful to endure and they wanted to stop. This point was reached at an average of only 350 microamperes by subjects who had been given the list of irrelevant symptoms and therefore attributed their feelings to the shock proper. In sharp contrast, it required an average of 1450 microamperes before this point was reached by subjects who attributed their symptoms to the pill rather than to the shock. Obviously, the attribution of symptoms has a major impact on the pain experience.

Because pain is notoriously manipulable, differences, even of this magnitude, may not be completely surprising. More revealing, perhaps, is some of the recent Russian work on interoceptive conditioning[11] which has demonstrated that even such presumably nonmalleable states as the feelings associated with micturition are astonishingly manipulable. Working with patients with urinary bladder fistulas, investigators have, by essentially cognitive procedures, been able to induce subjects with almost empty bladders to report an intense need to urinate, as well as to induce subjects with full bladders to report no particular urge to do so.

I trust that it is by now tediously clear that cognitive factors are, indeed, major determiners of the labels we attach to bodily states and of the affective tone we attribute to these states. There seems little need to elaborate or belabor the point further. For the remainder of this paper, I want to examine some of the implications of this way of thinking about bodily states, and to see how some very old biological phenomena look if we explicitly abandon the assumption of identity. Specifically, I would like to look into just one question: what happens if an individual makes a mistake; if, in the socially defined sense, he does not label a bodily state as most other people do.

If it is correct that the labels attached to feeling states are cognitively, situationally, or socially determined, it becomes a distinct possibility that an uncommon or inappropriate label can be attached to a feeling state. Where such is the case, we may anticipate behavior that appears bizarre and pathological. As an example of this possibility, consider the state of hunger. We are so accustomed to think of hunger as a primary motive, wired into the animal, and unmistakable in its cues, that even the possibility that an organism would be incapable of correctly labeling the state seems too far-fetched to credit. The physiological changes accompanying food deprivation seem distinct, identifiable, and invariant. Yet even a moment's consideration will make it clear that attaching the label "hunger" to this set of bodily feelings and behaving accordingly, is a learned, socially determined, cognitive act. Consider the neonate. Wholly at the mercy of its feelings, it screams when it is uncomfortable or in pain or frightened or hungry. Whether it is comforted, soothed, fondled, or fed has little to do with the state of its feelings, but depends entirely on the ability and willingness of its mother or nurse to recognize the proper cues. If she is experienced, she will comfort when the baby is frightened, soothe him when he is chafed, feed him when he is hungry, and so on. If inexperienced, her behavior may be completely inappropriate to the child's state. Most commonly, perhaps, the compassionate but bewildered mother will feed her child at any sign of distress.

It is precisely this state of affairs that the analyst Hilde Bruch[3] suggests is at the heart of chronic obesity. She describes such cases as characterized by confusion between intense emotional states and hunger. During childhood, she presumes, these patients have not been taught to discriminate between hunger and such states as fear, anger, and anxiety. If this is so, the patients may be labeling almost any state of arousal as hunger or, alternatively, labeling no internal state as hunger.

If Bruch's speculation is correct, it might be anticipated that the set of physiological symptoms considered characteristic of food deprivation are not labeled as "hunger" by the obese. In other words, the obese literally may not know when they are physiologically hungry. This may seem to be a remote possibility, but it appears to be the case. In an absorbing study, Stunkard,[19,21] has related gastric motility to self-reports of hunger in 37 obese and 37 normal-sized subjects. His experiment was simple and clear-cut. Subjects who had eaten no breakfast came to the laboratory at 9 a.m. and swallowed a gastric balloon. For the next four hours Stunkard continuously recorded gastric motility. Every fifteen minutes the subject was asked if he was hungry. He answered "yes" or "no," and that was all there was to the study. We have, then, a record of the extent to which a subject's self-report of hunger corresponds to his gastric motility. Let us note first that the two groups do not differ significantly in the extent of gastric motility, and second, that when the stomach is not contracting, obese and normal subjects are quite similar, both groups reporting hunger roughly 38 per cent of the time. When the stomach is contracting, however, the two groups differ markedly. For normals, self-report of hunger coincides with motility an average of 71 per cent of the time. For the obese, the coincidence is only 47.6 per cent. This difference is significant at considerably better than the .01 level of confidence.

Stunkard's work, then, would seem to indicate that obese and normal subjects do not refer to the same bodily state when they use the term hunger. Whether to interpret Stunkard's results as an instance of mislabeling, however, is still an open question. Stunkard himself tends to interpret his results in more psycho-dynamic terms, for he suggests denial mechanisms to account for at least some of his findings. In any case, it does seem that for the obese there is little correspondence between the bodily states commonly associated with hunger and the statement "I am hungry."

If all of this is correct, we should anticipate that if we were to directly manipulate gastric motility and the other symptoms that we associate with hunger, we should, for normals, directly manipulate feelings of hunger and eating behavior. For the obese, on the other hand, there should be no correspondence between the manipulated internal state and eating behavior. In order to test these expectations, Schachter, Goldman, and Gordon[14] did an experiment in which bodily state was manipulated by two means: first, by manipulating food deprivation so that some subjects entered an experimental eating situation with empty stomachs and others with full stomachs; second, by manipulating fear so that some subjects entered the eating situation badly frightened and others quite calm. Carlson[5] has indicated that fear inhibits gastric motility; Cannon[4] has demonstrated that the state of fear leads to the suppression of gastric movement and the liberation from the liver of sugar into the blood. Hypoglycemia and gastric contractions are generally considered the chief peripheral physiological correlates of food deprivation.

Our experiment was conducted within the framework of a study of taste. Subjects, all male undergraduates at Columbia, came to the laboratory in mid-afternoon or evening. On the previous evening they had all been asked not to eat the meal (lunch or dinner) preceding their experimental appointment. The experimenter's introductory patter was an expanded version of the following:

"A subject of considerable importance in psychology today is the interdependence of the basic human senses, that is, the way the stimulation of one sense affects another. To take a recent example, research has discovered that certain sounds act as very effective pain killers. Some dentists are, in fact, using these sounds instead of novocaine to 'block out' pain when they work on your teeth. Some psychologists believe that similar relationships exist for all the senses. The experiment we are working on now concerns the effect of tactile stimulation on the way things taste.

"The reason we asked you not to eat before coming here is that in any scientific experiment it is necessary that the subjects be as similar as possible in all relevant ways. As you probably know from your own experience, an important factor in determining how things taste is what you have recently eaten. For example, after eating any richly spiced food such as pizza, almost everything else tastes pretty bland."

The experimenter then manipulated pre-loading as follows:

To the Full Stomach subjects he said, "In order to guarantee that your recent taste experiences are entirely similar, we should now like you each to eat exactly the same thing. Just help yourself to the roast beef sandwiches on the table. Eat as much as you want — until you're full." The subjects spent about 15 minutes eating, and filled out a long food-preference questionnaire while they ate.

In the Empty Stomach condition, of course, the subjects were not fed. They simply spent the 15-minute period filling out the questionnaire about food.

Next, the subject was seated in front of five bowls of crackers and told, "Now that we are through with the preliminaries we can get to the main part of the experiment. We are going to have each of you taste five different kinds of crackers and tell us how they taste to you. These are very low-calorie crackers designed to resemble commercial products." The experimenter then presented the subject with a long set of rating scales and said, "We would like you to judge each cracker on each of the dimensions (salty, cheesy, garlicky, etc.) listed on this sheet. Taste as many or as few of the crackers of each type as you want in making your judgments; the important thing is that your ratings be as accurate as possible."

Before permitting the subjects to eat, the experimenter continued with the final stage of the experiment — the manipulation of fear. "As I mentioned before, our primary interest in this experiment is the effect of tactile stimulation on taste. Electric stimulation is the means we have chosen to excite your skin receptors. We use this method so that we can carefully control the amount of stimulation you receive."

For Low Fear, the subject was told, "In order to create the effect that we are interested in we need to use only the lowest level possible. At most, you will feel a slight tingle in your skin. Probably you will feel nothing at all. We are only interested in the effect of very weak stimulation."

For High Fear, the experimenter pointed to an eight foot high, jet-black console loaded with electrical junk and said, "That machine is the one we will be using. I am afraid that these shocks will be painful. In order for them to have any effect on your taste sensations, they must be of a rather high voltage. There will, of course, be no permanent damage. Do you have a heart condition?" The subject was then connected to the console by attaching a very large electrode to each ankle, and the experimenter concluded with, "The best way for us to test the effect of the tactile stimulation is to have you rate the crackers now, before the electric shock, to see how the crackers taste under normal circumstances, and then rate them again after the shock to see what changes in your ratings the shock has made."

The subject then proceeded to taste and rate crackers for 15 minutes. He was under the impression that he was tasting and, of course, we were simply counting the number of crackers he ate.* In this way we have a measure of the eating behavior of subjects who were empty or full and who were frightened or calm. Finally, of course, there were two groups of subjects—the obese, ranging from 14 per cent to 75 per cent overweight, and normals, ranging from 8 per cent underweight to 9 per cent overweight. We are co-varying three variables—pre-loading, fear, and obesity—in an eight-condition experiment. For expositional simplicity, I will present here only the main effects of this study, and will not give the data for all eight conditions.

To review expectations briefly: if it is correct that the obese do not label as hunger the bodily states of gastric motility and hypoglycemia, our several experimental manipulations should have no effects on the amount eaten by obese subjects. In sharp contrast, the eating behavior of normal subjects should directly parallel the effects of the manipulations on bodily state.

* It is a common belief among researchers in obesity that the sensitivity of their fat subjects makes it almost impossible to study their eating behavior experimentally—hence this roundabout way of measuring eating; the subjects in this study are taking a taste test, not eating.

Let us examine first the effects of pre-loading on the eating behavior of the two groups of subjects. From Figure 1 it will be a surprise to no one to learn that normals eat considerably fewer crackers when their stomachs are full than when they are empty. Fats stand in fascinating contrast. They eat as much — in fact slightly more — when their stomachs are full, as when they are empty (interaction p <.05). Obviously, the actual state of the stomach has nothing to do with the eating behavior of the obese.

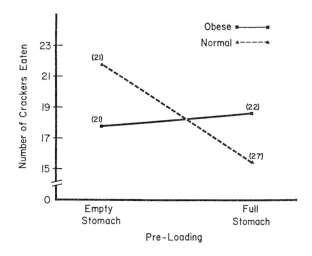

FIGURE 1 Effects of pre-loading on the eating behavior of normal and obese subjects. ( ) = number of subjects.

Turning to fear in Figure 2, we see much the same picture. High fear markedly decreases the number of crackers normal subjects eat and has no effect on the amount eaten by the obese (interaction p <.01). Again there is a small, although nonsignificant, reversal. The fearful obese eat slightly more than the calm obese.

There appears, then, to be little question that the obese do not label as "hunger" the same set of bodily symptoms as do normals. Whether we measure gastric motility, as in Stunkard's studies, or manipulate it, as I assume we have done in my studies, there is a high degree of correspondence between the state of the gut and the eating behavior of normals, and virtually no correspondence for the fat subjects.

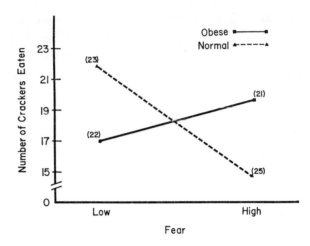

FIGURE 2 Effects of fear on the eating behavior of normal and obese subjects. ( ) = number of subjects.

Another way of summarizing these data would be: while almost anything we have manipulated experimentally has had a major effect on the amount eaten by normal subjects, apparently virtually nothing we have done has any substantial effect on the amount eaten by obese subjects. Still other indications of our complete inability to affect the eating behavior of obese subjects come from a small number of case studies I have been doing in order to compare the effects of injections of adrenalin and placebo on the eating behavior of hospitalized obese and normal patients. These studies are attempts to test the hypothesis offered some years ago[13] that the obese label the state of sympathetic activation as hunger — a suggestion that would lead to the expectation that the obese would eat considerably more when injected with adrenalin. This is not what we have found so far. Our obese subjects tend to eat very slightly more under adrenalin than under placebo (much as with the high and low fear comparison in Figure 2), while our normals tend to eat less with adrenalin in them. Again, in the work described so far, there is nothing that we have been able to do — from feeding to frightening to injecting with adrenalin — that has any effect on the eating behavior of the obese subject or that fails to have an effect on the eating behavior of the normal subject.

Keeping this in mind, let us turn to the work of the members of the Nutrition Clinic in St. Luke's Hospital in New York, chiefly Drs. Hashim and Van Itallie.[7] Summarizing their findings, virtually everything they do seems to have a major effect on the eating behavior of the obese and almost no effect on the eating behavior of the normal subject.

These researchers have prepared a bland and homogenized liquid diet, similar in taste and composition to the vanilla flavors of such commercial preparations as Nutrament or Metrecal, to which the subjects are restricted. They can eat as much or as little as they want of this relatively tasteless and uninteresting pap, but this and this alone is all they can eat for periods ranging from a week to several months. Some of their subjects get a large pitcher full of the stuff and can pour themselves a meal anytime they are so inclined. Other subjects are fed by a machine, which delivers a mouthful of the food every time the subject presses a button. Whichever feeding technique is used, the eating situation is characterized by three properties. First, the food itself is dull and unappealing. Second, eating is entirely self-determined — whether the subject eats, how much and when he eats is up to the subject and no one else. It should be specifically noted that absolutely no pressure is put on the subject to limit his consumption. Third, the eating situation is totally devoid of any social or domestic trappings. It is simply basic eating; it will keep the subject alive, but it's not much fun.

To date, six grossly obese and five normal subjects have been run in this set-up. Figure 3 plots the eating curves for a typical pair of subjects over a 21-day period. Both subjects were healthy, normal people who lived in and did not leave the hospital during the entire period of the study. The obese subject in this figure was a 52-year-old woman, 5′ 3″ tall, who weighed 307 pounds on admission. The normal subject was a 30-year-old male, 5′ 7″ tall, who weighed 132 pounds.

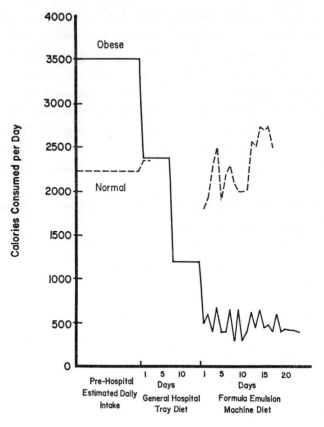

FIGURE 3 The effects of a formula emulsion diet on the eating behavior of an obese and a normal subject.

On the left of the Figure is shown the estimated daily caloric intake for each subject before entering the hospital — estimates based on detailed interviews. While in the hospital, but before entering the experimental regime, each subject was placed on a general hospital diet served on a tray. The obese subject was placed on a 2400-calorie general hospital diet for seven days and a 1200-calorie diet for the next eight days. As can be seen on the Figure, she consumed everything on her tray during this 15-day period. The normal subject was placed on a 2400-calorie general hospital diet for two days and he, too, ate everything on his tray.

With the beginning of the experiment proper, the difference in the amounts eaten by the two subjects becomes dramatic and startling. You will note immediately that the food consumed by the obese subject drops precipitously the moment she enters on the regime and remains at this incredibly low level for the duration of the experiment. This effect is so dramatic that one of the obese subjects who remained in the experiment for eight months dropped from 410 pounds to 190 pounds. The normal subject, on the other hand, drops slightly on the first two days, then returns to a fairly steady 2300 grams or so of food a day. These are typical curves. Every one of the six fat subjects has been characterized by this marked and persistent decrease in food consumption. Every one of the normal subjects has fairly steadily consumed about his normal amount of food.

Before worrying through possible interpretations of this data, I must note that there are certain marked differences between the two groups of subjects. Most important, the obese subjects have come to the clinic for help in their weight problems and are, of course, motivated to lose weight. The normal subjects are simply volunteers for an experiment. Without question, this difference could account for the effect, and until a group of obese volunteers who are unconcerned with their weight are run through the procedure we cannot be completely sure of the phenomenon. However, I would like to be sure that we do not, only on the grounds of methodological fastidiousness, dismiss these findings. As we have said, every obese subject was highly motivated to lose weight before entering the hospital, and certainly while in the hospital and before going on the formula emulsion diet. Yet, despite this motivation, no one of these subjects was capable of restricting his *home* diet successfully. When placed on the general hospital tray diet, motivated or not, every one of the obese subjects polished off his tray. Only when the food is dull and the act of eating self-initiated and devoid of any ritual trappings does the obese subject, motivated or not, severely limit his consumption.

On the one hand, then, we have a series of experiments that indicate virtually no relation between internal state and the eating behavior of the obese subject; on the other hand, this series of case studies seems to indicate a very close tie-up between the eating behavior of the obese and what might be called the circumstances of eating. When the food is uninspired and the eating situation uninteresting, the fat subject eats virtually nothing. The relationships are quite the reverse for the normal subject; his eating behavior seems directly linked to internal state but relatively unaffected by the external circumstances surrounding the eating routine and ritual.

Given this set of facts, it seems eminently clear that the eating behavior of obese and normal subjects is not triggered by the same set of bodily symptoms. Indeed, there is growing reason to suspect that the eating behavior of the obese is relatively unrelated to any internal gut state but is, in large part, under external control; that is, eating behavior is initiated and terminated by stimuli external to the organism.

Let me try to convey by a few examples what I mean by external control. A person whose eating behavior is under external control will stroll by a pastry shop, find the window irresistible and, whether or not he has recently eaten, will buy a goody. He will wander by a hamburger stand, smell and see the broiling meat, and although he may have eaten recently, will buy a hamburger. Obviously, such external factors — smell, sight, taste, what other people are doing, and so on — affect anyone's eating behavior to some extent. However, for normals, such external factors clearly interact with internal state. They may affect what, where, and how much the normal eats, but do so chiefly when he is in a state of physiological hunger. For the obese, I suggest, internal state is irrelevant and eating is determined largely by external factors.

Obviously, this hypothesis fits beautifully with the various data presented — as well it should, since it is an *ad hoc* construction specifically designed to fit the data. Let us see what independent support there is for the hypothesis and where it leads.

The essence of this notion of the external control of eating behavior is this: stimuli outside of the organism trigger eating behavior. In effect, since such internal states as gastric motility and hypoglycemia are not labeled as hunger, some cue outside the organism must tell it when it is hungry and when to eat. Of course, such cues are multiple, but one of the most intriguing is simply the passage of time.

Everyone "knows" that four to six hours after eating his last meal he should eat his next one. Everyone "knows" that within narrow limits there are set times to eat regular meals. In the absence of alternative cues, in the absence of competing alternatives to eating, the eating behavior of the externally controlled person should be time-bound. We should expect that if we manipulate time we should be able to manipulate the eating behavior of the obese subject. In order to do this, we[15] have simply taken two clocks and so gimmicked them that one runs at half normal speed and the other at roughly twice normal speed. A subject arrives at five in the afternoon, presumably to take part in an experiment on the relationship of base levels of autonomic reactivity to personality factors. He is ushered into a windowless room containing nothing but electronic equipment and a clock. Electrodes are put on the subject's wrists, his watch is removed so it will not be gummed up with electrode jelly, and he is connected to a polygraph. This consumes five minutes, and at 5:05 he is left completely alone with nothing to do for a true thirty minutes while presumably we are getting a record of resting-level rate of such autonomic indicators as galvanic skin response, cardiac rate, and so on. There are two conditions. In one the experimenter returns after a true thirty minutes and the clock reads 5:20. In the other, the clock reads 6:05 when the experimenter returns. In both cases, the experimenter is nibbling at crackers from a box as he comes into the room; he puts the box down, invites the subject to help himself, removes the subject's electrodes, and proceeds with the personality-testing phase of the study. For five minutes he administers a short version of the Embedded Figures Test. He then gives the subject a self-administering personality inventory and leaves him alone with the box of crackers for another true ten minutes. There are two groups of subjects — normal and obese — and of course the only datum we collect is the weight of the box of crackers before and after the subject has had a chance at it. If these ideas on internal and external controls of eating behavior are correct, we should anticipate the following pattern of results. Normal subjects, whose eating behavior is presumably linked to internal state, should be relatively unaffected by the manipulation and should eat roughly the same number of crackers whether the clock reads 5:20 or 6:05. The obese, on the other hand, if indeed they are under external control, should eat very few crackers when the clock reads 5:20 and a great many crackers when it reads 6:05.

The data of the experiment are presented in Figure 4, and indeed we do find that the obese eat almost twice as much when they think the time is 6:05 as they do when they believe it to be 5:20. For normal subjects, there is a distinct reverse trend (interaction $p=.002$)—a finding we had not originally anticipated, but one that seems embarrassingly simple to explain, as witness the several 6:05 normal subjects who politely refused the crackers, saying, "No, thanks, I don't want to spoil my dinner." Obviously, cognitive factors have affected the eating behavior of both normal and obese subjects with, however, a vast difference. While this cognitive manipulation of time serves to trigger or stimulate eating among the obese, it has the opposite effect on normal subjects, most of whom are at this hour, we presume, physiologically hungry, aware in the 6:05 condition that they will eat dinner very shortly, and unwilling to ruin their appetites by filling up on crackers.

In another study, Nisbett[10] has examined the effects of taste on eating behavior. He reasons that taste, like the sight or smell of food,

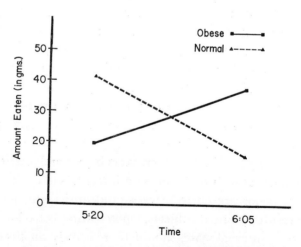

FIGURE 4   The effects of manipulated time on eating.

is essentially an external stimulus to eating. His experiment also extends the range of weight deviation by including a group of very skinny subjects, as well as obese and normal subjects. His purpose in so-doing was to examine the hypothesis that the relative potency of external vs. internal controls is a dimension directly related to the degree of overweight. If this is correct, it should be anticipated that the taste of food will have the greatest impact on the amount eaten by obese subjects and the least effect on skinny subjects. To test this, Nisbett had his subjects eat as much vanilla ice cream as they wanted. He gave them either a creamy and delicious, extremely expensive, preparation or an acrid brew of cheap vanilla ice cream and quinine, which he called "vanilla bitters." The effects of taste are presented in Figure 5, which plots the relation of a subject's ratings of how good or bad the ice cream was to the amount eaten. Obviously, when the ice

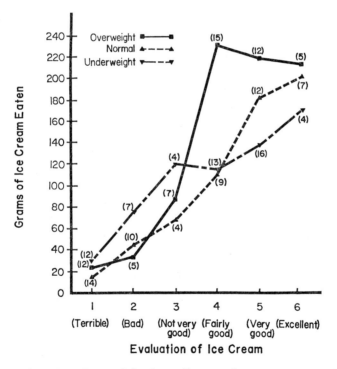

FIGURE 5 The effects of food quality on the amounts eaten by obese, normal, and skinny subjects. ( ) = number of subjects.

cream is rated as "fairly good" or better, obese subjects eat considerably more than do normal subjects, who, in turn, eat more than skinny subjects. When the ice cream is rated as "not very good" or worse, this ordering tends to reverse, with skinny subjects eating more than either normal or obese subjects. This experiment indicates that the external or, at least, the non-visceral taste cue does indeed have differential effects on the eating behavior of skinny, normal, and obese subjects.

The indications in this experiment—that the degree of dependence on external vs. internal cues is a dimension covarying with weight deviation—is a particularly fascinating finding, for if further work supports this hypothesis we do have the beginnings of a plausible explanation of why skinnies are skinny and fats are fat. We know from work such as that of Carlson[5] that gastric contractions cease after the introduction of only a small amount of food into the stomach. To the extent that such contractions are directly related to the hunger "experience," to the extent that a person is under internal control, he should literally "eat like a bird"—only enough to stop the contractions. Eating beyond this point should be a function of external cues—the taste, sight, and smell of food, perhaps the sheer joy of mastication. Externally controlled individuals, then, should have difficulty in stopping eating—a suggestion that may account for the notorious "binge" eating of the obese[20] or the monumental meals lovingly detailed by students (e.g., Beebe[1]) of the great, fat, gastronomic magnificos.

This last, loose attempt to account for why the obese are obese does in itself raise intriguing questions. For example, does the external control of eating behavior inevitably lead to obesity? I assume it is evident that logically such a link is not inevitable and that the condition of external control of eating behavior can indeed lead to the state of emaciation. An externally controlled person should eat and grow fat when food-related cues are abundant and when the person is fully aware of them. However, when such cues are absent or, for some reason, such as withdrawal or depression, the person is unaware of these cues, the person under external control should not eat, and if the condition persists, grow concentration-camp thin. If you go through the clinical literature you receive the impression that there is an odd but distinct relationship between extreme obesity and extreme emaciation. For example, 11 of 21 case studies in Bliss and Branch's book on anorexia nervosa[2] were at some point in their lives obese. In 8 of these 11 cases, anorexia was preceded and accompanied either by marked withdrawal or by an intense depression. In contrast, intense attacks of anxiety or nervousness (states that the experiment by Schachter, Goldman and Gordon[14] would suggest inhibit eating in normal subjects) seem to characterize the development of anorexia among most of the cases who were originally normal size.

At this point, obviously, these speculations are simply idea spinning — fun, but ephemeral. Let us return to the facts of the studies described so far. The results can be summarized quickly as follows:

1. Physiological correlates of food deprivation, such as gastric motility and hypoglycemia, are directly related to eating behavior and to the reported experience of hunger in normal-sized subjects but unrelated in obese subjects.[14,21]

2. External or non-visceral cues, such as smell, taste, the sight of other people eating, the passage of time, and so forth, affect eating behavior in obese subjects to a greater extent than in normal subjects.[7,10,15]

Given these basic facts, their implications ramify to almost any area involved in food and eating, and some of our recent studies have been concerned with the implications of these experimentally derived relationships for eating behavior in a variety of non-laboratory settings. Thus, Goldman, Jaffa, and Schachter[6] have studied the relationship of obesity to fasting on Yom Kippur, the Jewish Day of Atonement, which requires of the traditional Jew that he do without food or water for 24 hours. Reasoning that this occasion is one in which food-relevant external cues are particularly scarce, it seemed logical to expect that fat Jews would be more likely to fast than would normal Jews. In a study of 296 religious (defined as anyone who has been to a synagogue at least once during the past year for some reason other than a wedding or a bar mitzvah) Jewish college students, this does prove to be the case, for 83.3 per cent of fat Jews fasted as compared with 68.8 per cent of normal Jews who did so ($p < .05$).

Further, this external-internal control schema leads to the prediction that fat, fasting Jews who spend a great deal of time in synagogue on Yom Kippur will suffer less from fasting than fat, fasting Jews who spend little time in synagogue, and that there will be no such relationship for normal fasting Jews. It is apparent there will be far fewer food-related cues in the synagogue than on the street or at home. Therefore, the likelihood that the impulse for obese Jews to eat will be triggered is greater out of synagogue than in it. For normal Jews, this distinction is of less importance. In or out of synagogue, stomach pangs are stomach pangs. Again, the data support the expectation. Correlating the number of hours in synagogue against self ratings of fasting unpleasantness, there is, for obese subjects, a correlation of $-.50$, while for normal subjects the correlation is only $-.18$. Testing the difference between these correlations, $z=2.19$, which is significant at the .03 level. Obviously, for the obese, the more time in synagogue the less an ordeal is fasting. In contrast, for normals, the number of hours in synagogue has little to do with the difficulty of the fast.

In another study, Goldman, Jaffa, and Schachter[6] examined the relationship of obesity to choice of eating places. Generalizing from Nisbett's findings on taste, it seemed a plausible guess that the obese would be more drawn to good restaurants and more repelled by bad ones than would normal subjects. At Columbia, students have the option of eating in the University dining halls or in any of the swarm of more or less exotic restaurants, lunch counters, and delicatessens that surround this metropolitan campus. At this University, as probably at every similar place in the United States, student opinion of the institution's food is low. If a freshman elects to eat in a dormitory dining hall, he may, if he chooses, join a pre-pay food plan at the beginning of the school year. In so-doing, he prepays, at the rate of $16.25 a week for all of his meals. Any time after November 1, by paying a penalty of $15.00, the student may cancel his food contract. If we accept prevailing campus opinion as at all realistically based, we should anticipate that those for whom taste or food quality is most important will be most likely to let their food contracts expire. Obese freshmen, then, should be more likely to drop out of the food plan than normal freshmen. Again, the data support expectations, for 86.5 per cent of fat freshmen let their contracts drop as compared with the 67.1 per cent of normal students who did so ($p < .05$). Obesity does to some extent predict who chooses to subsist on institutional food.

In the final study in this series, Goldman, et al.,[6] examined the relationship of obesity to the difficulty involved in adjusting to new eating schedules imposed by time-zone changes. Thanks to the generosity of the Medical Department of Air France, it was possible to examine data of a study conducted by this department on medical effects of time-zone changes on 236 flight personnel regularly assigned to the Paris-New York and Paris-Montreal flights. This investigation was concerned largely with the effects of the east-to-west journey. Most of these flights are scheduled to leave Paris around noon, French time, fly for approximately eight hours, and land in North America sometime between 2:00 and 3:00 P.M., Eastern Standard Time. Flight-crew members all eat lunch shortly after takeoff and, being occupied with landing preparations and serving passenger needs, are not served another meal during the flight. They land in North America, then, some seven hours after their last meal, at a time that is past the local lunch hour and before the local dinner time.

The Air France study was not directly concerned with reports of hunger or eating behavior, but the interviewers systematically noted all individuals who volunteered that they "suffered from the discordance between their physiological state and meal time in America." (Quotation from letter from Dr. Lavernhe.) The interpretation of this coding is not completely clear-cut, but it appears to apply chiefly to fliers who complain that they either do without food or make do with a snack until local dinner time. Probably some complainers are those who eat a full meal on landing, are then sated at local dinner time, and are again physiologically hungry at a time long past local dinner time. In either case, it should be anticipated that the fatter fliers, sensitive to external rather than internal cues, should most readily adapt to local eating schedules and be least likely to complain of the discrepancy between American meal times and physiological state.

Given the rigorous physical requirements involved in air crew selection, there are, of course, relatively few really obese people in this sample. Nisbett's[10] experiment, however, has indicated that the relative reliance on external versus internal cues may well be a dimension covarying with the degree of weight deviation. It seems at least reasonable, then, to anticipate that even within a restricted sample there should be differences between the heavier and lighter members of the sample. This is the case. Comparing the 101 flying personnel who are overweight (0.1 per cent to 29 per cent overweight) with those 135 fliers who are not overweight (0 per cent to 25 per cent underweight) we find that 11.9 per cent of the overweight complain as compared with 25.3 per cent of the non-overweight ($p < .01$). It does appear that fatter flying Frenchmen are less likely to be troubled by the effects of time changes on eating.

Obviously, I make no pretense that the *only* explanation of the results of these last three studies lies in this external-internal control formulation of eating behavior. These studies were deliberately designed to test implications of the general schema in field settings. As with any field research, alternative explanations are legion, and within the context of any specific study are simply impossible to overrule. At the very least, however, these results are non-obvious and are consistent with this formulation.

The studies of Yom Kippur and of the eating behavior of Columbia freshmen and French fliers may at this point seem wildly remote from my introductory remarks, but they derive ultimately from this skepticism about the assumption of an identity between a physiological state and a psychological or behavioral event. The physiological correlates of food deprivation may or may not be associated with "hunger," may or may not be associated with eating. Explicitly abandoning the assumption of identity has for "hunger" proved experimentally worthwhile. I suspect that doing so for many of the other feeling states that are presumably physiologically well-anchored may prove equally rewarding.

Some Psychophysiological Considerations of the Relationship Between the Autonomic Nervous System and Behavior

MARVIN STEIN

This section of the conference is concerned with autonomic activity and emotion. Dr. Schachter's interesting presentation has raised several important questions about the relationship between the autonomic nervous system and behavior. First, what constitutes autonomic activity? Dr. Schachter utilized exogenous epinephrine as a means of manipulating the degree of sympathetic arousal. The use of epinephrine as a sympathomimetic agent can be traced historically to the notion that autonomic nerve impulses are chemically transmitted. In 1904 Elliott[8] suggested the possibility that when a sympathetic nerve impulse reached a target cell, epinephrine was released. The basis of this hypothesis was that epinephrine and sympathetic nerve activity were similar in their action, whether it was one of activation or inhibition. Similar ideas were applied to parasympathetic nerves and their comparison with the action of muscarine.[7] A series of studies supported the concept of the chemical transmission of autonomic nerve impulses.

Stimulation of the sympathetic nerves results in the release of a substance which accelerates the heart and has properties similar to epinephrine. Cannon and Uridil[3] found that stimulation of sympathetic nerves increased the rate of the denervated heart and raised the arterial blood pressure, but did not dilate the pupil. If the substance released following stimulation were epinephrine, dilation of the pupil would have been expected. This strange absence of effect on the iris was readily explained when it was recognized that norepineph-

---

MARVIN STEIN New York Hospital — Cornell Medical Center, New York. Present address: Downstate Medical Center, Brooklyn, New York

rine and not epinephrine was the mediator of adrenergic nerve action.[9,11] The present concept of sympathetic nervous system activity is that, after the appropriate stimulus, norepinephrine is secreted at all sympathetic endings, and simultaneously some epinephrine and norepinephrine are secreted by the adrenal medulla. Any stimulus that leads to the release of epinephrine and norepinephrine from the adrenal medulla also results in the simultaneous release of norepinephrine from sympathetic nerve axons.

Catecholamines other than norepinephrine have also been suggested as adrenergic nerve transmitters.[10] They include dopamine and isopropylnorepinephrine; however, the possibility of these substances serving as neurotransmitters requires additional investigation.

The catecholamines have many and varied actions in the body. They stimulate the heart, dilate the bronchi, inhibit smooth muscle of the intestine, and modify metabolic processes in various ways. Each catecholamine has a characteristic pattern of effects. The majority of the excitatory or stimulatory effects are blocked by specific agents such as ergotoxine and phenoxybenzamine. The inhibitory effects, the metabolic effects, and the stimulant effect on the heart are not blocked by these antagonists. This has led to the classification of adrenergic receptors into two major types.[1] Alpha receptors are those which are blocked by ergotoxine, dibenamine, phenoxybenzamine, etc., and beta receptors are those which are not blocked. Compounds are also available which selectively inhibit the beta receptors.[18,20] Norepinephrine combines primarily with alpha receptors, isopropylnorepinephrine chiefly with beta receptors, and epinephrine acts on both alpha and beta receptors.[28]

Research in recent years has clarified many of the problems related to the sympathetic nervous system and the metabolism and actions of sympathomimetic amines. It is a complex system, and apparently the injection of epinephrine cannot be considered as a means of producing autonomic or sympathetic arousal. It would seem preferable to refer, as Dr. Schachter did from time to time, to the state induced by exogenous epinephrine as one of physiological arousal.

Rothballer[21] has suggested that epinephrine has some central action, probably of an unspecific arousing or excitatory nature, which accounts for patients' frequent complaints of tremor, restlessness, and anticipation. It has been shown in unanesthetized cats that doses of epinephrine as small as 2 to 5 micrograms per kilogram of weight administered intravenously produce repeatable EEG activation and behavioral arousal in the naturally sleeping cat.[21] Responses to moderate intravenous doses of epinephrine and norepinephrine simulate many of the responses to electrical stimulation of the brain stem reticular formation.[21] These include EEG activation, reflexes involving the brain stem, and spinal motor facilitation. There is considerable evidence showing that epinephrine acts directly on the brain stem, specifically the mesencephalic reticular formation and posterior hypothalamus, to produce the effects in the spinal cord and in the EEG. The actual capacity of epinephrine to cross the blood-barrier is not known, and most likely is small. However, indirect evidence suggests that epinephrine can penetrate in sufficient quantity to exert some central effects.[21]

Dr. Schachter's epinephrine study may be interpreted, therefore, as one in which the subjects were aroused by the nonspecific central excitatory action of epinephrine. The design of the study, however, does not permit an evaluation of the physiological arousal following the administration of epinephrine as compared to the saline placebo. Dr. Schachter utilized only the pulse rate as the index of arousal and measured the pulse rate before the injection and immediately after the session with the stooge. In interpreting the results in relation to the arousal hypothesis, it would have been helpful to know if uninformed subjects were more physiologically aroused than the other groups prior to the attempt to induce an emotional state. The findings in the previously published report[22] suggest that the manipulation itself may have a specific effect, because the change in pulse rate was greater for the anger situation than for euphoria. Furthermore, as Dr. Schachter has written, even the injection of a placebo may produce arousal in some subjects. It would appear that superimposed on the nonspecific central effect of epinephrine was the subject's response to the experimental situation, as well as particular associations that the somatic symptoms may have evoked. With such a complex interaction of factors and without measures of arousal, it is difficult to know if there were or were not differences in the level of arousal for the various groups prior to the emotional manipulation. Continuous measures of pulse rate are very important in a study such as Dr. Schachter described, as epinephrine is removed from the circulation in a few moments[30] and any effects of the exogenous epinephrine may have subsided before the manipulation took place. There are no data available to deal with this possibility.

It would have been of considerable interest if several physiological parameters as indexes of arousal could have been recorded during the study. The need for multiple indicators of physiological arousal or autonomic activity has been frequently demonstrated. For example, we have found in our studies of airway caliber that peripherally accessible measures do not always provide a valid index of the underlying bronchiolar changes. A characteristic respiratory pattern found in asthma is one of shortened inspiration and prolonged expiration. This pattern has been demonstrated clinically and experimentally in humans and in experimental asthma in the guinea pig.[26] An identical respiratory pattern can be produced by stimulation of the vagus nerve[29] or by the response to a pain-fear stimulus.[23] In each of these conditions we have simultaneously measured airway resistance and compliance as indexes of airway caliber. Experimental asthma and vagal stimulation were both accompanied by bronchiolar obstruction, as indicated by significant increases in airway resistance and decreases in compliance. In contrast, the response to a pain-fear stimulus produced an identical respiratory pattern, but there was no evidence of bronchiolar obstruction. We have found that the asthmalike respiratory pattern following the pain-fear stimulus was related to the screeching of the animals. It was not observed when screeching was suppressed following tracheotomy.[23] This brief review of an aspect of our research emphasizes the need for careful consideration of physiological parameters.

To return more specifically to the question of arousal, no single physiological variable can be used to assign a level of arousal to a stimulus, nor to the responsiveness of an individual. A simple and convincing example of the problem presented by employing only one arbitrarily chosen variable as an index of arousal occurs when one variable shows a considerable response while a simultaneously measured variable shows no response. Lacey,[16] using the cold pressor test, has found that some cases showed cardiac acceleration but completely failed to show a palmar conductance response. On the other hand, some showed a response in palmar conductance, while the heart rate did not increase. In approximately 25 per cent of the subjects palmar conductance and heart rate were discordant. Lacey[16,17] and others[6] have also shown that the various autonomic indexes differ in response and in direction depending upon the type of situation. For example, Lacey[16] has reported that heart rate and palmar sweating both increase when a subject's feet are immersed in cold water or when the subject's task is mental arithmetic. In contrast, when male college students are shown pictures of nude females, heart rate decreases as palmar sweating increases.

The differential response to the situation has been clearly demonstrated by Lacey's investigation of a group of subjects exposed to a selected series of twelve stimulus situations. Four of the stimulus conditions and the responses to them will briefly be described as examples. 1) A visual attention task required the subject to note silently the patterns and colors produced by photic stimulation. 2) An empathic listening task caused the subject to listen to a 60-second dramatic reading of the thoughts and feelings of a dying man. 3) A mental arithmetic test comprised the thinking situation. 4) The cold pressor test was used as a test of withstanding pain. In all four situations there were increases in palmar conductance. The direction of the heart-rate change, however, differed. Cardiac deceleration occurred in the majority of the subjects in the visual attention task, while cardiac acceleration occurred in the majority during the cold pressor test. Time will not permit further discussion of this differential response to stimulus situations. However, it is worth noting that Lacey, on the basis of these observations and other findings, has related the direction of heart rate to the direction of attention and carotid sinus mechanisms. The over-all concept of arousal is open to serious question, and without the rigorous use of physiological indexes, and perhaps even multiple autonomic measures as indicators, it is difficult to evaluate studies using this concept.

Another serious question raised by Dr. Schachter's interpretation of his findings is that of the relationship between specific physiological responses and a given emotion. I do not wish to get involved in a lengthy discussion of the James-Lange theory of emotions. As Cantril and Hunt[14] commented in 1932, "The final formulations of his [James's] theory seem so vague and so open to personal interpretation that it appears almost useless for anything but the purpose of controversy." I would, therefore, like to direct my comments specifically to Dr. Schachter's study, and not to the James-Lange theory. There appears to be little, if any, question that cognitive factors play an important role in the labeling of somatic states. But Dr. Schachter's epinephrine experiment does not disprove nor prove the hypothesis or assumption of an identity between a physiological and emotional state. The study showed that following the injection of epinephrine a variety of emotions can be manifested. These include anger, euphoria, amusement, and fear. The results did not demonstrate that there were or were not specific physiological or biochemical differences in the specific emotions manifested. For the sake of this discussion, let us say that the epinephrine in the uninformed subjects produced arousal. The arousal plus specific cognitive input resulted in a specific emotion such as anger or euphoria. The response or specific emotion could be accompanied by specific physiological or chemical processes, or the response might even function as stimuli evoking specific bodily responses. These responses, in turn, might be involved in some type of feedback mechanism concerned with the original stimulus of epinephrine arousal. It is possible to conceive, for example, that different effects are selectively produced by the sympathetic nervous system in response to different types of stimuli. A differential response might be in terms of different amounts of catecholamines. Recently there has been considerable question about the ability of the adrenal medulla to secret selectively epinephrine or norepinephrine, depending upon the initial stimulus. There is data both to support and to refute this possibility.[12] The selective secretion of either of these substances, as pointed out by Hagen and Hagen,[13] will be expected to involve separate nervous pathways within the central nervous system and separate preganglionic fibers to different groups of epinephrine- and norepinephrine-secreting cells in the adrenal medulla. Different types of sensing organs would also be required. As more and more data become available about autonomic activity, we are growing closer to understanding emotions in terms of physiological and chemical processes. Finally, if arousal is concerned with the intensity, and not the qualitative aspect, of emotion one wonders if the intensity in humans could be manipulated by different dose levels of epinephrine. From a pharmacological point of view a single dose level is not adequate to construct a dose-response curve.

The thread running through the epinephrine and obesity studies is the assumption that cognitive factors play a major role in the labeling of bodily states. It has been suggested that obese subjects mislabel the internal state of hunger and respond primarily to external rather than internal cues. The strategy employed by Dr. Schachter consisted of attempts to manipulate gastric motility or hypoglycemia as indexes of hunger and to determine the effect on the eating behavior of normal and obese subjects. From a series of such studies it has been suggested that obese persons mislabel hunger.

Stunkard[27] has clearly demonstrated that there is a smaller correlation between the self-report of hunger and gastric motility in obese subjects than in normals. He has interpreted this finding as possibly being related to denial mechanisms. The findings of Van Itallie, as suggested by Dr. Schachter, may be the result of the motivation of the obese patient to lose weight and not to the mislabeling of an internal state. It is well known that obese individuals are sensitive about eating. It is interesting to note that in both the preloading study and the fear study the obese subjects ate fewer crackers initially than did the normal subjects. In fact, the intake or operant level of the obese subjects was about the same in all of the conditions in both studies. These data could also be interpreted as denial on the part of the obese subject in the experimental situation. Dr. Schachter did not measure hunger in terms of bodily state, such as gastric motility or hypoglycemia, so it is difficult to know if the obese subjects were mislabeling or if other factors were involved in the interesting differences found in their eating behavior.

Feeding behavior is complex, and involves appetite, hunger, satiety, and many other functions of the nervous system. Brobeck[2] has pointed out that although Cannon[4] and Carlson[5] considered gastric contractions as the source of hunger pangs, the mechanisms by which these sensations are perceived are not known. At the present time there is no hypothesis that explains how eating is regulated in a quantitative way on the basis of sensory experience. All of the major hypotheses concerned with the regulation of feeding consider the brain and, more specifically, the hypothalamus, as responding to what Sherrington[24] called "qualities of the circulating blood." Various factors have been suggested and include availability of glucose, metabolites related to the lipid reserves of the body, and thermal gradients. Various investigators emphasize one factor, while others believe that multiple factors are involved in the regulation of feeding. A multifactor concept of the regulation of feeding has been proposed by Brobeck.[2] Appetite is changed into satiety by eating and filling the stomach, by the alleviation of hypoglycemia, by the redistribution of body water, and by the thermal stress of specific dynamic action. It is important to recognize that even similar patterns of feeding activity may be the result of different factors, and without careful study it is difficult to know what process or processes obese subjects might be mislabeling.

Time will not permit a thorough discussion of the behavioral aspects of hunger, but I would like to emphasize again the need for multiple measures. Miller[19] has shown the value of using a diversity of rigorous behavioral tests in his experiments on hunger in the rat. He found, for example, differences in four measures of hunger in a study investigating the effects of food deprivation. The curve for stomach contractions followed that for volume of milk consumed. The other two measures, rate of bar pressing and the amount of quinine needed to stop the animal's drinking, had a different course. Miller concluded "that it is desirable to supplement the measures of amount of food consumed with other behavioral tests." Dr. Reuben Kron and

I[15] have found differential effects in several measures of nutrient sucking behavior in human infants as early as the first day of life.

Dr. Schachter's research has emphasized the role of cognitive factors in the labeling and mislabeling of bodily responses. We have observed what might be considered mislabeling in some patients with bronchial asthma. From some of our findings, it seems important not only to consider labeling, but also to attempt to obtain some information about the content of the label. This point perhaps can be illustrated by the following study. We have been investigating the experimental induction of asthmatic attacks utilizing emotional stimuli.[25] Airway resistance has been used as an index of bronchiolar obstruction. The situation in which the measurements took place was provocative of considerable tension. The patient was confined in a whole-body plethysmograph consisting of a small chamber locked from the outside by a door with the thickness of the usual bank vault. A complex array of electronic instruments was present in the plethysmograph and also filled the experimental room, which was visible to the patient through a small window in the wall of the box.

Although specific emotional stimuli could not be utilized in the early studies, several interesting observations were made. There was a wide variation in the patients' subjective impression of the situation and of their respiratory responses. An example was the response of a 32-year-old male with asthma and severe eczema of several years duration. The subject was placed in the allergen-free chamber, where airway resistance was recorded at appropriate intervals. The patient's initial comments were, "I'll be locked in the box, and what will happen if you two suddenly collapse and die?" The situation provoked a response of marked apprehension in the patient, accompanied by an increase in airway resistance and severe asthmatic symptoms. At the height of his asthmatic attack he complained, "This is the worst I have felt in weeks." At this point, an attempt was made to relieve the attack using suggestion and a nebulizer filled with saline. There was no significant change in airway resistance or in his symptoms. The asthma attack was promptly terminated when a bronchodilator was substituted for the saline. The experimental situation as an emotional event was strongly suggested by the findings, but the specific meaning of this emotional stimulus was not clear. Interviews immediately following the experiment provided little information. The emotional significance of being placed in the plethysmograph was not obtained until almost two years after the test situation. Then, during the course of psychotherapy, he revealed his feelings while in the "box," and a variety of fears, experiences, and attitudes about separation was brought out.

Another patient, a 43-year-old housewife with asthma of 25 years' duration, when placed in the same experimental situation, reported, "You doctors are wonderful. I think that this is grand!" Airway resistance measurements rose rapidly just moments before a frank asthmatic attack, in spite of her report that "This is the best I've felt in weeks." Further questioning and knowledge about the patient revealed that she had not mislabeled her bodily symptoms as it might appear, but that this was her characteristic way of relating in almost any situation. Her cognitive style was such that she always felt she was receiving some kind of help or support, even when an asthmatic attack had been induced.

The relationship between autonomic activity and emotions is complex and requires consideration of many factors, including physiological, cognitive, and social processes. The tendency to look for a simple stimulus response pattern in our understanding of behavior is gradually being abandoned. The organism cannot be separated from its environment, and upon closer examination, the distinction becomes nebulous. Dr. Schachter's presentation further emphasizes the importance of terminating the search for *the cause* and directing our energies toward a complete description of the organism in its environment, and the nature and the processes involved in behavior.

Inside Every Fat Man

NORMAN A. SCOTCH

"Inside every fat man," the old saw goes, "there is a thin man crying to get out." Now Dr. Schachter informs us that inside every fat man there is a crazy communication system. The system is so disturbed that all sorts of stimuli arising from internal body states having very little to do with hunger nevertheless bombard the fat man and deceive him into believing that he is hungry. Even when the fat man is full the message is the same—he is not full, he is hungry. Finally, external cues of all sorts remind the fat man, regardless of his internal state, that he is hungry.

My wife, who is not a scientist, has been making observations over the last decade that seem to parallel much of Dr. Schachter's work—in fact even antedate his work. She had noticed that, until a few years ago, if she sent me to the supermarket with a list of groceries *before* dinner, I was likely to return with all the items on the list plus a large additional number of "goodies" picked up at the gourmet shelf and the delicatessen counter. If she sent me immediately *after* dinner, I would get most of the items on the list and very little else. Since this latter system saved considerable money on her food budget, she soon tried to send me only after dinner. Of course, there were times when she needed things before dinner, and had to send me then. So we ate very well on the days following. The coda to this little anecdote is that my wife has noticed that as I have managed to put on an additional fifteen pounds over the last two or three years, it no longer matters when I shop—whether before or after dinner. I now bring home lots of goodies. As you might have guessed, she shops herself as much as possible.

The notion that a man's interpretation of his body state is as much a function of his psyche as it is of his physiology is really common-

---

NORMAN A. SCOTCH   School of Hygiene and Public Health, Johns Hopkins

place. We know, for example, that "crocks" (hypochondriacs) view every minor fluctuation in body state as life-threatening. We know furthermore that placebos often calm anxious patients by suggesting to them that powerful drugs are actually altering their body state, when in fact there is no powerful drug. On the other end of the continuum, a person's physical state frequently deteriorates so rapidly that death may be just around the corner without the person being the least aware of it. Moreover, we know that many people have heart attacks without their knowledge, and that these attacks are called *silent* infarcts.

Although we know all this — or think we do, anyway — Dr. Schachter has made a real contribution in the area: a lucid and fascinating account summarizing his excellent research. Without question, he is one of the very few to carry out systematic and inventive studies in a domain in which impressions rather than real knowledge have for so long dominated our thinking. Moreover, with work like that reported by Dr. Schachter we begin to accumulate some picture of the *mechanisms* and the processes involved in the interpretation of body state. This work, I believe, has very important implications for the clinical practice of medicine, as well as for certain types of research methodologies in the health field. I shall examine some of these briefly later.

Before turning from Dr. Schachter's actual research to the implications of his findings, I would like to make a few (perhaps gratuitous) comments on his methods and conceptualization. First, I am sure it was not meant for us to think that there are *not* highly stable body states that lead to uniform cognitive interpretations. In general, people undergoing torture feel pain, and people who are subjected to temperatures significantly below those to which they are accustomed feel cold, and so on. Second, although we are told that experimental subjects injected with epinephrine and exposed to spending time with either a euphoric or angry companion (in reality a shill for the experiment) tend to approximate the behavior of this companion, we are not told whether there is any variation in behavior. Do all the subjects in the anger experiment get equally angry? Do they get equally euphoric? Are there individual differences? If some get angrier than others, wouldn't it be valuable to be able to predict, on the basis of social-psychological-cultural variables, which of the subjects get most angry and which the least? Could not Dr. Schachter, using this same sort of refinement, tell us which of the obese subjects ate more crackers and which fewer? Finally, and along the same lines, although Dr. Schachter has been working mostly with hunger states and epinephrine injections, are there certain body states more amenable to psychic influence, and other states *less* amenable?

Dr. Schachter has been most informative while dealing with the psychological and personality aspects of the process in which he is interested. He has also attempted to approximate a social situation in his laboratory by providing the euphoric and angry shills. However, other than mentioning the social parameters of the problem, he has not explored the question of social and cultural influences on the interpretation of body state. One gets a sense from reading Schachter's paper that in order for perception to be distorted it is almost necessary to manipulate the organism powerfully in an experimental situation. I am sure this is not an impression that Dr. Schachter wishes to give.

Most social scientists would contend that there are important social and cultural parameters affecting the perception of body state. I would like to describe briefly several anthropological and sociological studies in this area that are congruent with the studies of Dr. Schachter and his associates.

In a sense, Dr. Schachter is telling us that there is no such thing as a "true body state." There is the body state as perceived by the individual when he is in one situation (say, alone) and the state perceived in another situation (say, with a euphoric companion). There is the body state as seen by the physician, and the state as measured by physiologists using various types of objective electrical and biological equipment. For a long time, anthropologists have been pointing out that there is no such thing as a "true physical environment," and that the habitat of a social group is perceived according to the culture of that group. The rainbow for some groups is interpreted positively, for others neutrally, and for still others with revulsion. And it is not simply a question of good or bad values placed on the same phenomena. They are actually perceived differently. Vernon Ray, an anthropologist, studied color perception among Northwest Coast Indians and found that the number of colors seen in the spectrum is a function of tribal membership: certain tribes see six basic colors, others seven, eight, or nine.[2,3] A study by Segall, Campbell, and Herskovits supports the notion of cultural influences on differential perception. In a recent book they show how people raised in an environment where there are virtually no rectangular or angular shapes respond differently to visual stimuli than do those raised in "angular" societies.[4]

If the external environment — the habitat — is perceived differently by different social groups, it comes as no surprise to learn that the internal environment is also perceived and interpreted in ways consistent with the culture or subculture. Consider, for example, Zborowski's minor classic, "Cultural Components in Response to Pain,"[5] in which the perception of pain is shown to vary with ethnic membership. Zborowski studied members of four different ethnic groups who were patients in a Veterans Administration hospital. The men were suffering from the same disease, so that presumably the pain stimuli were roughly the same for all the study participants. On the basis of multiple interviews with each subject, Zborowski concluded that Italians and Jews tend to perceive and express much more pain for the same physical condition than do Irish and Anglo-Saxon patients.

In a similar study, Irving Zola, attempting to get at the process of decision-making with regard to seeking medical care, found that different ethnic groups rely on different sanctions or "justifications" for going to the doctor.[6] Zola also found that different ethnic groups report a significantly different amount of presenting symptoms for the same diseases. (Italians most, Anglo-Saxons least, and Irish in between). Finally, and along the same lines, Koos in his study, *The Health of Regionville*,[1] learned that upper socio-economic people perceived and defined certain body states as encompassing possible symptoms of disease, whereas those from lower socio-economic groups saw the same kinds of body states as normal.

In all these studies, we are dealing with situations in which individuals in different social locations perceive as different what are presumably similar body states. The perception of body state is as much a function of social group membership as it is of personality factors. Not that I am suggesting that all members of particular social groups or social locations react uniformly to the same body states. I am simply restating what is well-known — that membership in particular cultures or subcultures, families, occupational groups, social classes, ethnic groups, geographic regions, or total societies, contributes to the ways in which we perceive and define our body states.

In turn, that perception affects in important ways *what we do about it*. Again, social and cultural factors tend both to set limits and to provide avenues for action. Thus, perception of hunger, regardless of "true" body state, sets in motion a complex set of behaviors designed to deal with hunger. This brings me to my concluding remarks. If it is true that perception is one of the first "acts" in a chain of acts, and that it is as pivotal an act as it appears, there are a number of important implications to be derived from Dr. Schachter's findings. Correct or incorrect perception of body states is the foundation for much of human behavior.

If a person perceives himself as being healthy when in fact he is ill, he might well act in ways that might hasten his death. He might continue to take part in activities that are very risky, for example, or he might fail to see a doctor in time, when medical intervention might save his life. On the other hand, if a person perceives himself as ill when he is well, there is an obvious loss of effectiveness in terms of "healthy" participation in day-to-day living. These are some dramatic implications of the consequences of distorted perception of body state, but less dramatic consequences are also of interest.

The practice of medicine today rests a good deal on verbal communication between doctor and patient. Almost any doctor taking a history implicitly acknowledges the possibility that the patient has distorted perception of his body state. But it would be helpful if doctors could be reminded explicitly that there is a systematic patterning of distorted perception that is a function of social and psychological characteristics.

Health surveys, such as the National Health Survey, provide us with a great deal of valuable information regarding almost every aspect of health and illness in the United States today. Investigators from several of these surveys have pointed out that there is considerable distortion in subjects' reporting, resulting from intent or from the passage of time. The kinds of distorted perceptions of body state reported by Dr. Schachter are probably reflected in reports on symptomatology in these surveys. The more we can understand about the process of distortion, the better we will be able to evaluate subjective self-reporting of such phenomena as symptomatology.

In closing, I would like to raise a few questions for consideration. I wonder about the implications of the work of Dr. Schachter and his colleagues for our understanding of the etiology of numerous diseases. Is there, for example, a possible relationship between certain aspects of the so-called autoimmune diseases and distorted perception of body states? Is it possible that distorted body perception may indeed contribute to the development of disease? To what extent can the possibilities of influencing perception of body state be utilized in therapeutic settings? I think that, in addition to the possibilities of manipulation and exploitation of people by influencing their perception, there is the opportunity for intervening and preventing certain unfortunate events from occurring.

Stimulation in Infancy, Emotional Reactivity, and Exploratory Behavior

V. H. DENENBERG

DISCUSSION

Biology and the Emotions
J. P. Scott   190

Analysis of Infant Stimulation
John W. M. Whiting   201

How stimulation and experiences during one's early life affect one's later behavior is of general concern to all of us. This is reflected in adages such as "Spare the rod and spoil the child," and "As the twig is bent, the tree's inclined." Many parents are concerned about protecting their child from, or exposing their child to, certain stimulus events during different stages in development. Our interests in our children's education, starting as low as the nursery school level, also reflect our belief that the stimulation to which they are exposed and the experiences they are acquiring will have important consequences in later life. And all of us know enough Freudian psychology to be interested in and concerned about the effects of certain sorts of early experiences upon personality development.

In spite of the general interest in, and the importance of, this topic, there had been almost no experimental research upon it until the early 1950's. Since then, a number of researchers have carried out experiments using animals. Some of the experiments have been de-

---

v. h. denenberg   Purdue University

signed to try to get an answer to a question derived from observations on humans, while others have been concerned with broader biological principles.

For the past ten years I have been interested in the general problem of the impact of stimulation during early life upon subsequent behavioral and physiological activities. The rat, mouse, and rabbit have been the research animals used in these experiments. I should like to describe some of our findings, together with those of some other experimenters, in the hope that I can convey to you a broad picture of the research activity in part of this growing field.

To keep within manageable boundaries, I shall restrict myself in the main to describing studies employing rats, although a wide variety of other species has been investigated as well. Furthermore, I shall deal primarily with the behavioral level, with only an occasional reference to physiological research. I have been asked to discuss the effects of infantile stimulation upon emotional behavior, and that imposes another restriction upon my comments. However, I plan to deviate a bit, and will bring in some research findings on exploratory behavior as well.

First, I will discuss the construct of emotional reactivity. Next will come a brief summary of research findings, in which I shall follow a developmental sequence — starting *in utero,* then to a discussion of the effects of manipulating physical and social stimuli between birth and weaning (the period of infancy), and then to a consideration of some postweaning phenomena. After this, I will describe some of our current activities. I shall conclude by making an abrupt phylogenetic shift to the human level to summarize some research findings concerning the effects of stimulation in the early life of the child. There appear to be interesting parallels between some of the rat data and the human data that ought to merit our consideration.

THE CONSTRUCT OF EMOTIONAL REACTIVITY

The rationale underlying the measurement of the construct of emotional reactivity in the rat is relatively simple. Like many other animals, the rat typically "freezes" when exposed to a set of strange and noxious stimuli. In addition, the animal is likely to urinate and/or defecate. Therefore, an "emotional" animal is defined, on the behavioral level, as one which exhibits little or no movement and which defecates or urinates in response to certain stimuli. Defecation is typically measured in one of two ways: the occurrence or nonoccurrence of the event can be noted, or quantitative data may be obtained by recording the number of boluses dropped in a standard unit of time. The typical measure of urination is in terms of whether the event occurred. There are several ways of measuring an animal's movement. Two common ones are latency to initiate movement following presentation of a noxious stimulus (which will cause the animals to "freeze"), and total activity in an open-field apparatus.

PROCEDURES FOR MEASURING EMOTIONAL REACTIVITY

Open-field Behavior

An open field is simply a relatively large arena illuminated with bright lights. Because of its size, strangeness, and bright illumination, rats will usually exhibit defecation, urination, and freezing behavior in such an apparatus. In our laboratory we use a field 45 inches square; the floor is marked off into 25 nine-inch squares for ease of counting locomotor activity. We usually give our animals a daily three-minute test for four successive days. We record the number of squares entered and the number of boluses defecated.

A circular open field is also commonly used. For example, Broadhurst[6] uses a field which is $32\tfrac{3}{4}$ inches in diameter. The floor is marked off into areas roughly equal in size. In addition to the usual bright lights, Broadhurst employs a white-noise stimulus of 78 decibels to increase the noxiousness of the situation.

Consummatory Behavior

As indicated above, a necessary component to induce emotional behavior is the presentation of some noxious stimuli. These can be external events, like the strange box and bright lights in the open-field test, or they may be internal. If an animal has always had food and water available *ad libitum,* then the internal stimuli arising from a period of water deprivation should constitute a set of novel and noxious events that may be expected to inhibit consummatory behavior when the animal is given an opportunity to drink. If this is true, the more emotional the animal, the greater should be his latency to drink and the less should be the amount of water he consumes in a test situation.

Emergence-from-Cage Behavior

Another common measure of emotional reactivity is the timidity test. In one version, rats are deprived of food for a period of time and then are given the opportunity to leave their home cage and venture out on an open alleyway to obtain food. The common measures of emotionality are latency to emerge from the home cage and distance traveled on the alleyway. Presumably the less emotional animal will leave the security of the home cage sooner and venture out a greater distance in his quest for food.

Behavior in a Learning Situation

In studying learning, the animal is generally placed in a novel stimulus chamber. If one is studying escape or avoidance learning, the reinforcing stimulus is one that is noxious to the subject. Thus, it may be possible to obtain measures of emotional reactivity to the strange apparatus, or to the reinforcing stimulus, or both. Common measures would be amount of activity during habituation training, and occurrence of defecation or urination during habituation or during testing.

Ratings of Emotional Reactivity

Another method of measuring emotionality is to make ratings of the animal's response to a disturbing stimulus. Often this is done when the experimenter tries to remove the animal from its home cage.

Let me now describe some of the experimental findings relating stimulation in infancy to the subsequent emotional reactivity of the rat. I have reviewed this material before,[9,10] so I will cover it here only briefly.

STIMULATION BEFORE BIRTH

Thompson[41] has shown that one of the determinants of the offspring's emotional reactivity is the presence of "anxiety" during pregnancy. An experimental group of rats was made anxious during pregnancy by being placed in a conditioning apparatus where they were presented with a buzzer. This buzzer had previously been associated with electric shock, and Thompson assumed that the presentation of the buzzer would be sufficient to elicit an "anxiety" response on the part of the female. The pups were tested at 30 days and retested at 130 days of age. An open-field test and a timidity test were used to assess emotionality. Pups born of mothers that had been made anxious during pregnancy were found to be more emotional than those from control mothers. Since Thompson's original paper, there have been a number of replications of this general phenomenon.[2,26,35]

The obverse of Thompson's findings have been reported by Ader and Conklin,[3] who were able to reduce the emotional reactivity of rat pups by handling the female three times during pregnancy. Open-field and timidity tests, administered at 45 and 100 days of age, indicated that the pups born of handled mothers were less emotional.

A third approach to the investigation of prenatal influences is that of Denenberg and Whimbey.[21] We generated high- or low-emotional mothers by the experimental procedure of handling or not disturbing females in infancy. When adult, these females were bred; at birth, cross-fostering was done within and between high- and low-emotional mothers, thus allowing us to separate the prenatal (but nongenetic) effects from the postnatal effects, and to keep both of these factors independent of the fostering effect per se. The offspring's weaning weight, and open-field activity and defecation scores were found to be significantly influenced by the experiences that their mothers had as infants. It will be convenient to summarize these findings in detail in a later section of this paper.

STIMULATION BETWEEN BIRTH AND WEANING: EFFECTS OF PHYSICAL STIMULI

Handling

The most widely used procedure for manipulating rats in infancy has been "handling." The pups are removed from the home cage (some experimenters leave the mother in the cage, while others remove her before removing the pups), placed in some sort of container (one partially filled with shavings, grid floor, wooden box) where they remain for a brief period of time (two to three minutes has been the modal time, although some experimenters have used intervals of eight minutes or longer), and then returned to the home cage. This procedure is usually administered once a day. The number of days of handling has ranged from one to the total time between birth and weaning (many experimenters wean their subjects at 21 days, although some do so earlier and others later).

The relation between handling in infancy and later emotionality has been well-established: handled animals are less emotional than nondisturbed controls, and these differences may last for almost the whole of the animal's lifetime. This phenomenon has been documented in several laboratories using different behavioral measures of emotional reactivity, including the open field;[15-18,20,43] activity and defecation measures in an avoidance learning situation,[30,43] consummatory behavior;[31,32] a measure of timidity when emerging from the home cage;[27] and ratings of emotionality.[27,43]

The long-lasting nature of the effects of handling in infancy has been shown in several studies. We have found significant differences in emotionality to be present at 180 days,[15] 218 days,[20] and 233 days[43] as a function of stimulation in infancy. The clearest evidence of the permanancy of these effects has been documented by Hunt and Otis,[27] who tested different groups of animals starting at 317 days, 322 days, or 544 days of age. In all instances, the nonhandled controls were

Electric Shock

found to be more timid than those that had been handled in infancy.

Shock, like handling, when administered prior to weaning, brings about a reduction in emotionality. Animals shocked on Days 11 through 20 were more active and showed a smaller per cent of defecating than did the nondisturbed controls,[20] while rats shocked on Days 1 and 2 of life emerged into an open field more quickly and were more active than controls when tested at weaning.[11] Using the consummatory index of emotionality, Levine[31,32] showed that rats shocked in infancy consumed more water following 18 hours of thirst than did nondisturbed controls. In a similar vein, Lindholm[34] has reported that rats shocked for the first 10 days, the second 10 days, or the first 20 days of life had a shorter latency to drink and consumed more water than nondisturbed controls following 24 hours of water deprivation.

Sensory Restriction

An intriguing experiment concerning the consequences of temporary sensory deprivation was reported by Wolf in 1943.[48] One group of rats was deprived of hearing from Day 10 to Day 25 in life, a second group was deprived of visual stimulation from Day 12 to Day 25, and a control group was not disturbed. As adults, the animals were trained to run after food upon presentation of a visual or an auditory signal. No differences were found among the three groups. The crucial part of the experiment was a competition test for food between pairs of animals, one of which had been deprived of auditory experience while the other had been deprived of visual experience (controls were randomly paired). When a visual signal was used as a cue stimulus, the animals deprived of audition in infancy won more often; these same animals lost more often when the cue was an auditory signal. The controls that won in one situation tended to win in the other situation. Although there was no direct measure of emotional reactivity, the nature of the test suggests that the competition engendered a state of emotionality that interacted with the animal's prior deprivation experience. The major findings of Wolf's study have been verified by Gauron and Becker.[22]

STIMULATION BETWEEN BIRTH AND WEANING: SOCIAL INTERACTIONS

In addition to studying the effects of handling and other forms of physical stimulation upon later behavioral and physiological processes, we have also been carrying out studies in which we have varied the social context within which rats have been reared. Our interest here has been to try to set up analogs to investigate hypotheses derived from human observations in a clinical context. One problem with which we have been concerned is that of the relationship between the "anxiety" (i.e., emotionality level) of the mother and the subsequent emotional reactivity of her offspring. Our second concerned the relationship between "multiple mothering" and offspring emotionality.

In our first study,[19] we used an experimental technique to induce emotional upset in mothers. Starting the day after birth, experimental mothers were removed from their nest cage, placed on a grid, and given 10 shock trials of .2 milliamperes, each trial lasting for 11 seconds. They were then returned to their pups. To manipulate the multiple mothering variable, a litter of pups was cared for alternately by its own mother for one 24-hour period and a different mother for the second 24 hours. This procedure was carried out daily until weaning. Open-field testing in adulthood revealed that maternal anxiety and multiple mothering both led to a significant increase in emotional defecation.

In our second experiment,[38] we again used the procedure of rotating mothers between two litters as our multiple mothering technique. However, we used an individual difference measure to manipulate maternal emotionality rather than our previous procedure of shocking mothers during the nursing period. We tested all our young, mature, nonpregnant females in the open field and classified them into three groups based upon their activity and defecation scores. Low-emotional females were those that had high activity and low defecation scores; high-emotional females were those with the opposite pattern; medium-emotional females were, of course, intermediate. The rotation variable and the maternal emotionality variable were combined in a 2 x 3 factorial design.

Analysis of the offspring's open-field behavior at 50 days of age revealed that multiple-mothered pups were significantly less active than control pups. In addition, the more emotional the mother the less active her pups. There is some ambiguity concerning the latter finding, because the relationship between maternal emotionality and offspring emotionality could be caused by a genetic factor, a prenatal event, or by the postnatal behavior of the mother toward her pups. Two subsequent experiments isolated a prenatal genetic component and a postnatal maternal component, each of which independently affected later open-field activity.

When two such divergent techniques as shocking mothers during the nursing period and classifying nonpregnant females on the individual difference measure of open-field activity yield similar results with respect to their offsprings' emotional behavior, one has considerable confidence that the phenomenon has some degree of generality. Therefore, the next logical step (we thought) was to generate, by experimental means, females that were more or less emotional (as measured by open-field behavior). By this procedure we would eliminate the genetic component, which is part of any individual difference measure, and at the same time would avoid all the difficulties involved in removing mothers from cages to shock them, as in our first procedure.

Our technique for experimentally generating high- and low-emotional females was our usual handling procedure. Litters were handled for the first 20 days of life, and the females were then set aside to be used as breeders. A large number of nonhandled females also were set aside to be used as breeders. These animals were not used in any experiments. The sole purpose for their being was to serve as experimental mothers. When they were sexually mature, they were bred, and we then examined the behavior of their offspring. We did obtain significant differences in offspring behavior, but they were definitely not in the predictable direction.

In our first experiment[21] we cross-fostered between handled and nonhandled mothers, thus allowing us to separate any prenatal, nongenetic component from the postnatal component. As a further control, we did not foster a number of litters. Our first measure was body weight at weaning. We found that pups reared by mothers that had been handled in their infancy weighed significantly more than pups reared by mothers that had not been so disturbed. This difference disappeared by the time the subjects were 53 days old. Our next set of measures, and the key ones with respect to our prior research, concerned open-field performance between 50 and 53 days of age. First of all, we found that subjects reared by mothers handled in infancy defecated significantly *more* in the open field than did subjects reared by nonhandled mothers. In other words, being reared by a less emotional mother brought about a significant increase in emotional reactivity of the pups. Clearly, this is quite contrary to our prior findings concerning maternal emotionality and offspring emotionality.[19,38] Our activity data were, if anything, more unexpected: a significant prenatal-postnatal interaction was obtained. Young born of mothers not handled in infancy and reared by mothers that had been handled were significantly more active than the other groups in the experiment. As part of another experiment (which will be described later in this paper) we have repeated part of the prior study and have obtained results consistent with our original findings.[43]

At the present time, we have no explanation for this seeming paradox concerning the relationship between maternal and offspring emotionality. Several experiments now in progress may help cast some light upon the phenomenon. One point that emerges from these data, however, deserves emphasis: the experiences of the mother during her infancy have significant effects upon her offsprings' weaning weight and open-field behavior. Thus, the impact of infantile stimulation extends into the next generation. This "second generation effect" is via a nongenetic mechanism, so the question arises as to how much modification of the offsprings' behavior and physiology can be induced by manipulation of the mother's experiences during her early life. Another question concerns the effects upon the third or fourth generations, and so on.

STIMULATION AFTER WEANING

Social Deprivation

Several studies have found that isolation housing after weaning acts to enhance emotional reactivity. In one such study, group-reared rats were found to explore more, crouch less, and squeal less than isolation-reared animals.[40] In another study, Ader, Kreutner, and Jacobs[4] found that group-reared rats were more active and took less time to emerge from their home cage. Recently Ader[1] has reported an additional confirmation of his findings.

Social Interaction in an Enriched Environment

We have seen that rats which have not received some form of extrinsic stimulation in infancy, such as handling or electric shock, are more emotional than those which have received extra stimulation. An important theoretical question is whether this heightened emotionality can be reduced by some form of postweaning experience. In a broader context, this question is concerned with the "reversibility" of the effects of early experience. One possible reason why the effects of early experience have been found to persist throughout most of the animal's lifetime[27] is that they have not had an opportunity to gain other experiences that may act to modify emotionality. The usual procedure, at the time of weaning, is to place rats in relatively small laboratory cages where they can eat and drink *ad lib*. They can also move around somewhat, but that is about all they can do. Such an environment would not be expected to aid an animal in reducing its emotional behavior. On the other hand, if animals were reared in a complex environment, which allowed them to engage in a broader range of social interactions, to develop their perceptual capacities more fully, and to interact with a wide range of environmental stimuli, such experiences might be beneficial in reducing emotional reactivity. This suggests that free environment experience may be of "therapeutic" value for animals reared undisturbed between birth and weaning.

To test this hypothesis, we carried out a study in which handled and nonhandled rats were given 25 days of experience in a free environment immediately after weaning, while other handled and non-handled animals spent the equivalent time in the usual laboratory cages.[15] The free-environment animals were also placed in laboratory cages at the end of their 25-day period. All animals were tested in the open field at 180 days of age. As usual, handled animals were found to be less emotional than nondisturbed controls, as measured by open-field defecation. But the important finding was that those groups which spent 25 days in a free environment immediately after weaning were also significantly less emotional than the groups reared entirely in laboratory cages after weaning. We have confirmed these findings in a second study;[16] in addition, we failed to find any evidence that the presence or absence of social interaction between the sexes affected emotionality.

A THEORY CONCERNING THE EFFECTS OF PHYSICAL STIMULATION IN INFANCY

The literature reviewed above indicates that handling or electric shock, when administered some time between birth and weaning, results in a reduction of emotional reactivity. I have proposed two hypotheses relating the effects of physical stimulation in infancy to adult performance measures.[9] The first hypothesis is that emotional reactivity is reduced as a monotonic function of amount of stimulus input in infancy. The theoretical curve relating that stimulus input to adult emotionality is shown in Figure 1.

The second hypothesis is concerned with the relationship between emotional reactivity and adult performance. In a performance task that contains some form of noxious element (e.g., learning with shock reinforcement), it is reasonable to expect there will be an optimal level of emotionality for efficient performance. As one moves away from this optimal level, task performance should drop off, resulting in an inverted-U function. But such a function would not be expected to obtain at all levels of task difficulty. For example, given a very simple task (like straightaway behavior), one expects performance to improve as emotionality increases, assuming that the more emotional animal is more motivated; on the other hand, the least emotional subjects should have the best performance when the task is very diffi-

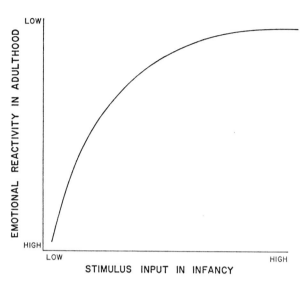

FIGURE 1 Theoretical curve relating stimulus input in infancy to emotionality in adulthood. (From Denenberg[9])

cult.[5,28] Only under conditions of "moderate" emotionality should the inverted-U function appear. This is another way of describing the Yerkes-Dodson Law,[5] which states that the optimal level of motivation for a task decreases as task difficulty increases. The postulated relationship between performance in adulthood and emotionality level for tasks of varying degrees of difficulty is shown in Figure 2.

Throughout this discussion we have been making a basic assumption, but one which has not been explicitly stated. The assumption is that there is, indeed, a construct which can be called "emotional reactivity" and that the various operational procedures which have been previously described all measure this construct, at least to a fair extent. Although there is a goodly body of data consistent with such an assumption, direct evidence concerning its validity is lacking. One way to test the validity is to give a group of animals a variety of behavioral tests that include a number of presumed measures of emotional reactivity, obtain the intercorrelations, and then factor-analyze the intercorrelation matrix. If a factor of "emotional reactivity" emerges, it may be taken as strong support for the validity of the as-

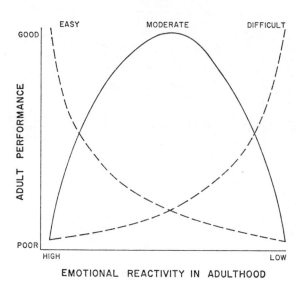

FIGURE 2 Theoretical relationship between performance in adulthood and emotionality level for tasks of varying degrees of difficulty. (From Denenberg[9])

sumption. We have recently completed such a study,[43] but before I can describe it I must first describe some of our current research activities, so that the factor-analysis study can be placed in its proper context.

SOME CURRENT RESEARCH ACTIVITIES

It seems reasonable to assume that a person's behavior at any moment in time is determined, to a significant extent, by his accumulated past experiences. Often it is not one isolated experience in a person's background that is the critical determiner of his action, but a particular combination of experiences. This has been the general idea underlying some of our current research activities.

Much of the research done to date on the effects of early experience has used the design of giving independent groups of animals different sorts of stimulation during a particular time interval and then studying the consequences of that stimulation in later life. A logical extension is to give animals different experiences at successive time intervals and to study how these experiences act, singly and in combination, to affect the animal's later behavior. One can, in other words, experimentally program life histories for animals by scheduling particular experiences to occur at specified times in the animal's life span. We have chosen to manipulate two classes of variables in our research — stress and social variables.

The design of one such experiment is shown in Table I. Sixteen different programs of life experiences were generated by combining four independent variables in a 2^4 factorial design. These variables were sequenced to include the time span from conception to 42 days of age.

The first variable concerned the mothers of our experimental subjects. These females had either been handled during their infancy or were nondisturbed controls; they had received no experimental experience during their lifetime.[21] The second variable was our usual handling procedure administered between birth and weaning. The last two variables concerned housing conditions. The mothers gave birth to their litters and reared them to weaning in our standard maternity cage or in a free environment.[36] After weaning the Ss were housed in standard laboratory cages or in free environments.[15,16] From 42 days until adult testing, all animals remained in standard laboratory cages.

This experiment, and several others like it, were set up in factorial designs so that analysis of variance procedures could be applied. We were particularly interested in the interactions, because these could give us some inkling on how experiences at different times in ontogeny affected each other.

However, there is another way of conceptualizing this experiment. The 16 different groups resulting from all possible combinations of the four variables listed in Table I may be conceived as 16 different experimentally created individuals. The variance among them is a function of known experimental manipulations during the animal's early life history. The nature of the design is such that the usual genetic contribution to individual difference is eliminated as a systematic source of variance. Now, the study of individual differences is generally carried out by psychometricians rather than experimental

TABLE I

Design showing how 16 groups of experimental animals are formed by varying four independent variables over the time span from conception to 42 days of age

Conception to Day 21	Day 1 to Day 20	Day 1 to Day 21	Day 21 to Day 42	Day 42 to Day 233
Experience of natural mothers in their infancy	Handling experience of pups	Preweaning housing	Postweaning housing	
Nonhandled	Nonhandled	Laboratory cage	Laboratory cage	All animals reared under standardized living conditions. Animals were given a series of behavioral tests from Day 220 through Day 233.
			Free environment	
		Free environment	Laboratory cage	
			Free environment	
	Handled	Laboratory cage	Laboratory cage	
			Free environment	
		Free environment	Laboratory cage	
			Free environment	
Handled	Nonhandled	Laboratory cage	Laboratory cage	
			Free environment	
		Free environment	Laboratory cage	
			Free environment	
	Handled	Laboratory cage	Laboratory cage	
			Free environment	
		Free environment	Laboratory cage	
			Free environment	

psychologists, but there is nothing to prevent the experimental psychologist from employing the tools of the psychometrician to help analyze data. Here the techniques of correlation and factor analysis were employed.

We can now return to the question that ended the last section of the paper: Is there any evidence for the validity of the construct of "emotional reactivity"? By administering to the 16 groups in Table I a battery of tests, many of which are presumed to measure emotionality, and factor-analyzing the resulting intercorrelation matrix, it is possible to determine whether a factor of emotional reactivity is obtained. If such a factor is found, not only does it verify the assumption concerning the existence of such a construct; it also establishes that an animal's early experiences are crucial in determining its level of emotionality, as the only systematic differences among the 16 groups in the experiment just described are those brought about by the four independent variables.

In the experiment depicted in Table I, three males and three females from different litters were randomly assigned to each of the 16 experimental groups. Beginning at 220 days of age, each animal went through 14 days of testing, during which 44 behavioral measures were obtained. These included one set of measures of avoidance learning, a variety of measures of emotional reactivity and exploratory behavior, and several other variables.

In addition, two behavioral measures on the litter-mates of the Ss used in this study were also included — one of avoidance learning and one of open-field activity. This experiment was part of a larger study in which the litter-mates were tested starting at 70 days of age. Since experimental treatments were assigned to whole litters, the litter-mates had received the same experimental experience as had the subjects in this study. For purposes of this paper it is not necessary to specify all the tests used. (See Whimbey[43] for complete details.)

The intercorrelations among these 46 variables across the 16 groups were obtained, and the matrix was factor analyzed. This factor analysis allowed us to eliminate redundant variables and to reduce the number of variables to 23. These were factor-analyzed anew, and the resulting six factors were submitted to an orthogonal rotation. Three of the six factors could be clearly identified as "emotional reactivity," "field exploration," and "consumption-elimination."[44,45] Only the first two factors are relevant to this discussion. Table II summarizes those measures which loaded on one or both of these factors.

TABLE II

Behavioral measures and factor loadings for the factors of emotional reactivity and field exploration

Age in days when measure obtained	Variable	Emotional reactivity factor	Field exploration factor
220	Emotionality rating	.697	.309
220–223	Open-field		
	Activity	−.028	−.840
	Defecation	.394	.436
224	Novel stimulus box		
	Time in stimulus half	−.529	−.188
	Number of crossings	−.418	−.636
225	Social stimulus box		
	Time in stimulus half	−.051	.234
	Number of crossings	−.228	−.777
226–230	Avoidance learning		
	Number of avoidances	.175	.005
	Defecation on Day 230	.743	.039
231	Consummatory behavior during day		
	Food	−.905	−.123
	Water	−.493	.012
233	Open-field defecation	.504	.424
70–73 *	Open-field activity	−.330	−.651

* This measure was obtained on litter-mates of the animals used in this study.

For the factor called emotional reactivity there are positive loadings on the experimenter's rating of emotionality when the animal was first removed from its home cage at 220 days of age, and on three defecation measures obtained in the open field or in an avoidance learning apparatus. Food and water consumption the day following termination of avoidance learning have negative loadings, as does total open-field activity between 70 and 73 days of age. This latter finding is particularly relevant, because these data were obtained from the litter-mates of the subjects used in this experiment.

Two other measures also correlated with this factor are the time score and crossings score for the novel stimulus apparatus. A word about this is in order. We had designed this apparatus in the hope that it would induce "curiosity" in the rat. The unit was a rectangular box 20 inches long by 13 inches wide. One half was empty, while the other half contained a bell and pieces of rubber, wood, metal, and wire. In addition, a child's pinwheel, turned by air from an air hose, was nailed to the back of the latter half. Our two measures were the amount of time spent in the stimulus half, and the number of crossings between the two halves of the box. The object was to have the novel stimuli and the moving pinwheel induce exploratory behavior, but observations of the animal's behavior in this unit indicated that the noise from the hose and pinwheel caused the animals to run away and freeze, rather than to explore. Therefore, behavior in this apparatus is more a reflection of "fear" than it is of curiosity, which is consistent with the negative loadings found on the emotional reactivity factor.

The second orthogonal factor, that of field exploration, is characterized primarily by negative loadings on two activity measures in the open field, number of crossings in the novel stimulus box, and number of crossings in the social stimulus box. (This was a unit identical with the novel stimulus apparatus, except that a tethered rat of the same sex was placed in one compartment in lieu of the novel stimulus.) Defecation in the open field loads moderately on this factor with an opposite sign.

The measure of learning — total number of avoidance responses made over the five days of testing — is included in the analysis to show that it is independent of these two factors.

These data supply good evidence for the validity of the construct of emotional reactivity. In addition, the factor analysis has allowed us to isolate a field exploration factor. The next question is: what effect did each of our early experience variables have upon these two dimensions of behavior? That question can be answered by examining the analysis of variance findings for each test that had significant factor loadings on either of the two factors.

Early Experience Variables Affecting Emotional Reactivity

The procedure of handling pups in infancy was the most powerful variable affecting emotional reactivity. Handled subjects were rated as less emotional, they defecated less in the open field and in the avoidance learning apparatus on the last day of learning, and they were more active in the novel stimulus unit.

A second variable affecting emotionality was free-environment experience postweaning. This experience reduced emotional reactivity, as measured by open-field defecation, thus adding confirmation to our previously reported findings.[15,16]

Surprisingly, the characteristics of the mother were not found to have any significant main effect upon the several measures of emotionality, although this variable was involved in a few significant interactions. The mother's experience had been expected to have a strong effect, because we had reported that pups reared by females which had not been handled in their infancy defecated significantly less than pups reared by handled mothers.[21] In that experiment the subjects were tested in the open field between 50 and 53 days of age.

Early Experience Variables Affecting Field Exploration

The two conditions that significantly affected exploratory behavior were the nature of the mother and the pups' handling experience in infancy. Pups reared by nonhandled mothers were more active in the open field during the first four days of testing and on the 14-day retest, and their latency to leave the starting square in the open field was less than that of pups reared by handled mothers.

On the other hand, handling resulted in heightened open-field activity during the first four days of testing, shorter latency to leave the starting square in the field, and greater activity in the novel stimulus unit and the social stimulus apparatus.

Although neither the preweaning nor the postweaning housing conditions were expressed as significant main effects, they were present in a large number of significant interactions.

Effects of Programed Life Experience
Upon Emotional Reactivity and Exploratory Behavior

We may now return to the question that initiated much of this research: how do various experimental programs of life experiences affect the animal's subsequent behavior? The answer lies in the analysis of the interactions obtained in our many analyses of variance. A brief summary is as follows: few significant interactions were obtained in the analyses of variance of the variables that correlated with the factor of emotional reactivity. A tentative conclusion is that the particular sequence of experiences is not relevant for this dimension of behavior. On the other hand, a very large number of significant interactions were obtained among the four independent variables when the tests measuring exploratory behavior were analyzed. From this we may conclude that the particular combinations and sequences of life experiences play a significant role in affecting exploratory behavior.

OTHER DATA RELATING HANDLING
TO EXPLORATORY BEHAVIOR

The factor analysis findings just described revealed that handling in infancy affected emotionality and exploratory behavior in adulthood. It also revealed that these two behavioral dimensions were orthogonal to each other. Independent confirmation has been obtained in two other sets of experiments in which very different methodologies have been used to evaluate the effects of infantile handling.

The first experiment[12] employed handled and nonhandled male rats. After weaning, some subjects in each of these two major groups were reared in isolation, others were reared in social groups, and still others were reared in social groups until 90 days of age, at which time they were placed in isolation. Testing started at 100 days. The experimental rat was placed in the empty stem of a T-shaped unit. The left chamber of the unit was designated as the "social chamber." It contained a wire-mesh screen in one corner, behind which was a male rat as a stimulus object. The right chamber, called the "novelty chamber," contained three white, wooden, three-dimensional, randomly cut-out forms, which were attached to the walls of the chamber by means of wire. The forms were small enough for the subject to be able to manipulate them with his paws and to chew them. To maintain novelty, the forms were changed daily. The animal was left in the chamber for 10 minutes; the amount of time spent in each of the three chambers was recorded, as was total activity and defecation. The subjects were tested for four successive days. As compared to nonhandled controls, rats that had been handled in infancy spent significantly more time in the novel and social chambers. These rats were also significantly more active than were the controls. Thus, the greater time in the social and novel chambers may reflect greater activity, rather than any intrinsic interest in the stimuli provided. To test this hypothesis, a covariance analysis was employed to determine whether the temporal differences between handled and nonhandled animals would hold up when the effects of activity were held constant (statistically). Handled rats still spent significantly more time in the novel chamber than did the controls, but there was no significant difference between the two groups with respect to time in the social chamber. These results indicate that novelty seeking behavior is affected by handling in infancy, and is independent of gross activity.

We have recently completed a set of experiments investigating tactual and visual exploratory behavior of handled and nonhandled rats.[8] The test apparatus was a unit constructed in the form of a Greek cross. Variation in tactual stimulation was manipulated by varying the characteristics of the unit's floor.[8a] Four floors were constructed; these could be rank-ordered along a dimension of tactual stimulus variation. One floor was smooth Masonite and represented our minimum stimulus variation condition. A second floor was completely covered with a sandpaper of intermediate coarseness; this represented a somewhat greater amount of stimulus variation. The third floor had four outer chambers with the sandpaper of intermediate coarseness, and a center chamber of Masonite. Thus, the rat, in crossing from one outer chamber to another, had to cross from sandpaper to smooth Masonite, to sandpaper. The fourth floor was like the third, in that it was sandpapered in the outer chambers and smooth in the center, but the sandpaper was of four different grades, ranging from quite smooth to very coarse. This represented the greatest degree of stimulus variation. Animals that had been handled in infancy and nonhandled controls were given four days of testing in these units. Summed over all four stimulus units, handled subjects entered significantly more chambers than did nonhandled controls, and they defecated significantly less than the controls. However, the key part of the experiment is the interaction of Infantile Stimulation and Stimulus Condition. That interaction was found to be significant, and is graphed in Figure 3. As the degree of stimulus variation increased (from smooth Masonite to four grades of sandpaper plus Masonite), the exploratory behavior of handled rats also increased, but the exploratory behavior of control rats was depressed.

In a second experiment,[8b] handled and nonhandled rats were allowed to explore the Greek cross under one of three brightness conditions: 1) all chambers black; 2) all chambers white; or 3) two of the four peripheral chambers black, the other two white, and the center chamber gray. Again the interaction was significant, and the curves (Figure 4) are quite similar to those seen in Figure 3.

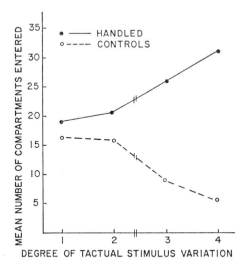

FIGURE 3 Exploratory behavior as a function of degree of tactual stimulus variation for handled and nonhandled rats.

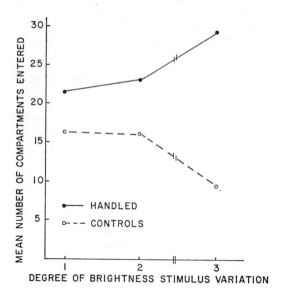

FIGURE 4 Exploratory behavior as a function of degree of brightness stimulus variation for handled and nonhandled rats.

The final experiment[8b] in this series was to allow handled and nonhandled rats to roam around the Greek cross under conditions of 1) small differences in illumination in the various compartments, or 2) large differences in illumination. The significant interaction is shown in Figure 5.

In these various experiments, rats have been presented novel, three-dimensional visual stimuli, social stimuli, tactual stimuli, and stimuli differing in brightness; they have been tested in open fields, curiosity boxes, a T-unit, and an apparatus with a Greek cross configuration. In each instance we have been able to separate the exploratory behavior from the emotional reactivity dimension by one of several procedures (factor analysis, covariance analysis, or the particular experimental procedure employed). In each instance we have found significant relationships between handling in infancy and later exploratory behavior. We may now conclude that handling rats in infancy increases exploratory behavior as measured by a number of different procedures and, independent of this dimension, handling also reduces emotional reactivity as measured in a wide variety of test situations.

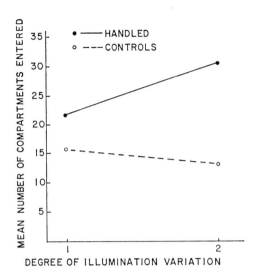

FIGURE 5   Exploratory behavior as a function of degree of illumination variation for handled and nonhandled rats.

INFANTILE STIMULATION WITH THE HUMAN

I should like to conclude this discussion by summarizing briefly several early experience studies with human infants. Two studies have employed procedures somewhat similar to the animal work on handling.[7,46] White and Castle worked with a population of infants reared in an institution where they normally receive minimal handling. The standard hospital procedure defined the treatment received by their control group of babies (N=18) during the first month of life. Over and above this standard hospital procedure, their experimental babies (N=10) received an extra 20 minutes of daily handling (i.e., tactile stimulation) for 30 consecutive days, starting with the sixth day of life. No differences were found between the groups with respect to the development of prehension, over-all development as measured by the Gesell Schedules, rate of weight gain, or general health. During the handling period, the infant's eyes were blindfolded. This was to prevent them from obtaining extra visual experience, as the researchers also planned to measure visual behavior. Between one month seven days and three months twenty-one days of age both groups of babies were given a number of tests. The experimental group was found to engage in significantly more visual exploration of its environment than did the controls.

Casler[7] also studied institutionalized infants. He used sixteen babies and employed tactile stimulation as his independent variable. Matched pairs were created by matching the babies on Gesell scores, chronological age, length of time in the institution, and age at the time of admission to the institution. At the beginning of the study the mean age for the experimental and control group was 22.5 weeks, with a range from 4 to 59 weeks. The experimental subjects received a total of 1000 minutes of added tactile stimulation (20 minutes of extra stimulation per day, five days a week, over a 10-week period). At the end of the 10-week period the children were retested on the Gesell Schedules. The psychologist who had tested the babies originally also administered the retest; she did not know the nature of the experiment nor which babies were in which groups. The absolute developmental quotients declined for both the experimental and control babies over the 10-week interval. However, the control group declined significantly more than the experimental group on the three Gesell subtests measuring adaptive development, language development, and personal-social development; they also declined significantly more on the total developmental quotient. The two groups did not differ on the motor development schedule. Thus, the effects of the extra infantile stimulation was to prevent the experimental babies from having as great a loss in developmental rate as the control babies. Casler suggests that the absolute decline in developmental quotients may be attributed to the over-all lack of stimulation in the babies.

A significant anthropological contribution has been made to the early experience literature by Landauer and Whiting.[29] They were familiar with the findings that handling, shocking, shaking, or other forms of "stressful" stimulation in animal infancy leads to a more rapid rate of physical growth.[13,14,33] They selected height of adult males as their index of physical growth and attempted to relate this measure, cross-culturally, to apparently stressful infant care practices. In their major analyses, two classes of stressers were used—piercing and molding. Examples of the former are "piercing the nose, lips, or ear to receive an ornament; circumcision, inoculation, scarification, or cauterization." Molding included "stretching the arms or legs, or shaping the head (usually for cosmetic purposes)." In their first sample, Landauer and Whiting found that those societies that stressed infants during the first two years of life produced males 2.49 inches taller in adulthood than did societies where such stress experiences were absent. (In this analysis, each society represents one case.) In an independent cross-validation, the authors examined 30 additional societies. Males from societies that practiced piercing or molding during the first two years of life were 2.64 inches taller than males from societies where such practices were absent. The mean difference is highly significant in both studies. The authors carried out further analyses which showed that geographical-genetic differences and amount of sunlight (inversely measured via amount of annual rainfall) could reasonably be ruled out as alternative explanations of the height difference.

In a subsequent report Whiting[47] examined the relationship between stress during the first two weeks of life and onset of menarche. In those societies in which the infant was subjected either to pain or shaping during the first two weeks, the median menarcheal age was 12.25 years; in societies where these events did not occur, the median age was 14.0 years. These findings provide an interesting parallel for the study by Morton, Denenberg, and Zarrow,[37] who found that sexual maturation occurred earlier in both male and female rats if they had been handled in infancy.

Let me conclude this section by citing two studies that implicate prenatal experiences as important determiners of postnatal behavior in the human. Two of the studies described previously[21,38] presented evidence that the emotionality level of the rat was affected by two prenatal mechanisms, one presumably genetic[38] and the other nongenetic and presumably uterine.[21] In an extension of the Ottinger, et al.,[38] experiment, Ottinger and Simmons[39] asked the question: is there any evidence that the emotional behavior of the neonate is affected by the mother's prenatal emotionality level? Maternal emotionality was measured by a paper-and-pencil anxiety test administered to expectant mothers during each trimester of pregnancy. The ten women with the highest scores and the nine with the lowest scores were classified as high- and low-anxious, respectively. The neonate's emotionality level was measured by a special apparatus that recorded crying and gross bodily activity. These behaviors were measured for 30 minutes before the infants were fed, and for 30 minutes after feeding. Testing started within 24 hours after the infants were born and continued for three successive days. When measured before feeding, the babies of the high-anxiety mothers were found to cry significantly more than the babies of low-anxiety mothers. No significant difference in crying was found between the two groups after feeding. Although not significant, Ottinger and Simmons report that the body movement data are similar in pattern to the crying behavior. The mechanism underlying this difference is not known, but the results do show that behavior within the first 24 hours after birth can be predicted from knowledge of the mother's prenatal anxiety level.

The last set of experiments involves the technique of abdominal decompression. This is a procedure developed by Heyns,[23] a South African physician, to aid expectant mothers during labor. The patient puts on an airtight suit with plastic spacer. This unit fits around the abdomen. Attached to it is an air evacuation pump, which can be controlled by the patient. During labor, the air pressure can be lowered around the abdominal region, causing the fetus to rise in the body cavity. The patient starts the decompression as soon as she becomes aware of the onset of a labor contraction. This technique has been found to reduce labor pains and to result in a significant shortening of time in labor. (A dome-shaped plastic blister that can be fitted easily around the patient's abdomen is now available, and eliminates many of the disadvantages of the suit. See Wajdowicz[42] for a picture of the apparatus, as well as for her personal description of the experience of giving birth with the aid of abdominal decompression.)

In addition to being beneficial for the mother, Heyns proposes that abdominal decompression is also beneficial to the fetus. He argues that there is poor oxygenation to the fetus during the last trimester of pregnancy, and that abdominal decompression should improve the fetus's oxygen supply. To test this hypothesis, Heyns and his colleagues[24,25] obtained Gesell developmental quotients on babies of white and Bantu mothers who had received a number of abdominal decompressions during the last trimester of pregnancy. Control babies were also tested. The white and Bantu control babies had developmental quotients of 105.9 and 110.2, respectively. The babies of two experimental groups of decompressed white mothers had mean developmental quotients of 128.1 and 130.1, and the one experimental Bantu group had a mean developmental quotient of 134.1. The mean difference between the experimental and control babies is highly significant. In addition to the t-tests between groups, correlation coefficients were computed to determine if a relationship existed within groups between number of decompressions during pregnancy and the child's developmental quotient. For one group of white women a correlation of .34 was obtained (N=162), while a correlation of .63 was obtained for the Bantu women (N=16). Both correlations are significantly different from zero.

SUMMARY

Research with the rat has clearly shown that stimulation during different stages in ontogeny (prenatally, during infancy, after weaning) can have significant effects upon the animal's adult behavior and can, indeed, carry over and influence the offspring of stimulated females. Two of the major dimensions affected by stimulation in early life are emotional reactivity and exploratory behavior.

I have described one of our major current research activities, which is concerned with experimental programing of life experiences. This program has two objectives. One is to determine how experiences at different stages in ontogeny interact with each other in affecting various behavioral and physiological characteristics in adulthood. Second is to view the psychometric problem of individual differences from an experimental framework of controlled experiential inputs during development. In a real sense we are now able to produce, by experimental means, a wide variety of "personalities" among our experimental animals.

Finally, I discussed several studies with humans, in which stimulation during the prenatal period or during infancy was found to affect a variety of measures, including visual exploratory behavior, developmental quotients, crying, and adult height. These latter findings should not be taken to mean that our animal work has been confirmed on the human level, or that we can extract principles from the animal research that may be applied directly to humans. What it does mean, however, is that findings on the human level appear to parallel the findings that have been obtained on the animal level, thereby suggesting that the principles underlying the effects of stimulation in early life may have broad and far-reaching biological implications.

Biology and the Emotions

J. P. SCOTT

A reader of these papers could gather the impression that there is much controversy and disagreement among their authors. Actually, there is a great deal of orderliness in the facts that have been presented, and I shall comment not only upon Denenberg's paper but will show that most of the other papers fit into a general biological framework of ideas. Most of the apparent disagreement arises out of overgeneralization.

The Concept of Adaptive Function

I shall begin with the general biological concept of adaptive function. There are at least five different possible functions of the behavioral and physiological reactions that are labeled emotional.

J. P. SCOTT Bowling Green State University

1 Communication. The back-arching, spitting, hair-lifting, and tail-switching of an angry cat has the obvious function of communicating with other individuals of the same species. This is what Darwin meant when he spoke of the expression of emotion. I have begun some preliminary work along these lines, and started by asking some trained student actors to express common emotional sounds indicating fear, pain, etc. I have not yet analyzed these completely, but one of the obvious results is an enormous amount of variation among individuals. This raises the question: how can one individual recognize the emotional signals of another? People find it difficult to recognize their own emotions, as Schachter's evidence shows, and should find it even more difficult to recognize the emotions of others.

2 The magnification and prolongation of reactions to behavioral stimulation. This is very important in connection with agonistic behavior, as the emotions of fear and rage tend to be long-lasting, but relatively unimportant in the orientation reaction and other expressions of investigatory behavior, where emotional reactions tend to be confined to a momentary response.

3 The primary stimulation of behavior. The physiological changes associated with hunger and thirst contribute strongly to consummatory or ingestive behavior, but such primary physiological changes appear to be very unimportant in agonistic behavior.

4 The reinforcement of behavior. This may have either positive or negative effects depending upon the emotion involved. Because emotional reactions are in many cases aroused slowly (although not in exploratory behavior), the reinforcing effect may be associated with quite irrelevant behavior and result in the strengthening of inappropriate responses. This is a key concept in analyzing the relationship of emotions to maladaptive behavior. One of the problems every human being faces as he grows up is that of understanding and managing his emotional responses.

5 The elicitation of the physiological preparatory and reparatory processes associated with stress. These were well-described by Dr. Brady and the two discussants of his paper, and require no further comment.

It should be clear by now that the investigators represented in this book have been discussing different sorts of emotional function. In addition, they have been discussing behavioral reactions and situations that have very different general functions. These can be determined easily from Table I.

TABLE I
Systems of behavior and associated emotions

Behavioral system	General function	Associated emotional responses
Ingestive	Nutrition	Hunger, thirst
Investigative	Information gathering	Relief of fear?
Shelter-seeking (comfort-seeking)	Finding optimum environment	Discomfort
Sexual	Reproduction	Sexual enjoyment
Epimeletic (care-giving)	Care of young, species mates	Parental enjoyment?
Et-epimeletic (care soliciting)	Signaling for care and attention	Distress, discomfort, fear, etc.
Allelomimetic	Group cohesion, safety	Fear (of being alone)
Agonistic	Dispersal	Anger, fear (of attack)
Eliminative	Disposal of urine and feces; secondarily, signaling	Relief (bladder, bowels)

The Concept of Systems

I will now comment upon the proceedings of this conference in relation to another general biological concept, that of systems. This is closely related to the concept of function, and indeed a system may be defined as a group of entities organized in relation to a common function. Table I includes a list of the important behavioral systems that have been identified in animals, each system being composed of several alternate behavior patterns with a common general function. A behavior pattern, in turn, is a piece of behavior which has a special function. Not only is the externally observable behavior of an animal organized around general functions; associated with each of these is a group of physiological and emotional responses that also form part of the system.

The Table immediately illustrates that there are large blank spaces in our knowledge concerning the emotional basis of behavior. Why, for example, has no one worked with the emotional basis of epimeletic (care-giving) behavior, and particularly with that of maternal behavior? It is also obvious that the content of this conference is extremely limited. In the first paper, Karl Pribram gives a lucid account of the neurophysiology of exploratory or investigative behavior. This is followed by presentations on the physiology of agonistic behavior, and Dr. Schachter's interesting experiments on agonistic behavior, which he has now extended to ingestive, or consummatory, behavior. Dr. Denenberg's developmental approach to the problem is concerned with exploratory, agonistic, and ingestive, or consummatory, behavior. Consequently, we have considered emotions from only three of these general systems, and have omitted the rest. While the behaviors discussed in this volume are certainly highly important, why did no one report studies on sexual emotions?

Some of the disagreements among the papers of this series arise because people are working with different organizational systems. There is no real disagreement between the "Centralist" and the "Peripheralist," as Cannon and now Schachter have demonstrated. The sensations from viscera are relatively unimportant in agonistic behavior, but assume much larger significance in behavior directly concerning the viscera, as in ingestive behavior. Pribram lays great stress on equilibrium and homeostasis in connection with investigative behavior, and indeed there is considerable evidence that in agonistic behavior a state of balance exists between various parts of the cerebral cortex and the hypothalamus. However, equilibrium is disturbed for much longer periods in the latter sorts of emotional states. And, as Pribram has pointed out, the general theories of emotion are not mutually exclusive. There is something in the James-Lange theory, but it is not complete and inclusive, universally applicable to all emotional state and to all behavioral systems.

It is a great pleasure to be asked to discuss Dr. Denenberg's fine piece of research, as it relieves me from any responsibility for criticism and, rather, permits me to enlarge and speculate on what has been said. There has been much criticism lately of experimental psychologists who use elegant techniques to attack inconsequential problems. Dr. Denenberg cannot be included in this group, for he has made use of the most sophisticated experimental and statistical techniques to attack one of the basic problems of education and clinical psychology — the effects of early experience.

THE CONSTRUCT OF EMOTIONALITY

I would like to point out that as the construct of emotionality has developed, it has tended to be extremely limited and consequently to be both behaviorally and physiologically incomplete. To demonstrate, I would like to put this construct and its associated behavior into the context of a rat as a biological organism, fitted by a long process of evolution to survive — first as a close associate of man in his dwellings and agricultural pursuits, and second as an even closer associate of the psychologist in his laboratory.

The behavior of defecating in a strange place, which Calvin Hall[3] first used as an index of emotionality, presumably serves some adaptive function for the individual organism or for the social group as a whole. It appears to be related to a phenomenon which occurs generally in the animal kingdom — that of localization in the sense of forming an emotional attachment to a particular place. When a rat is forcibly removed from his familiar home cage he does two things: he increases his rate of urination and defecation, and he begins to explore the new area. The latter has an obvious adaptive function, as eventually the rat will either find his way back home or will become familiar with the new environment. The eliminative behavior has a less obvious usefulness. By analogy with human reactions, it indicates that the animal has become fearful. Since it will be relieved in familiar surroundings, this fear tends to make him stay in familiar areas.

In a more natural situation, elimination would serve to give more obvious clues to those predators that hunt by scent, and to this extent the behavior appears maladaptive. Many small rodents that live in burrows have regular latrine areas not accessible to predators. It has been suggested that urination acts as a territorial marker, but there is no good evidence to support this. Rats show no tendency to avoid areas that have been used by others for elimination. Further, one would expect that an animal would mark a familiar area rather than an unfamiliar one. Another suggestion is that urination and defecation simply make the new area smell familiar. Finally, there is the possibility that the rat uses its own fecal boli as a sort of blazed trail for finding its way home again. If none of these suggestions is correct, we would have to assume that this behavior is simply a severe emotional disturbance, with no function other than to reinforce the unpleasantness of being in an unfamiliar locality.

From the physiological viewpoint, the construct of emotionality is also incomplete. Obviously, there are many different kinds of emotions associated with different and often conflicting systems of behavior, and there is no reason to suppose that all of them can be described by one measure.

One would assume that emotional elimination in the rat would be a part and a symptom of a more general fear reaction to strange places, and that, as in other animals, it would be associated with important changes in the hypothalamus, if not the direct result of them. By using several different measures of emotional behavior, Denenberg and his co-workers have come to an interesting conclusion. The rate of defecation appears in two different factors and has a relatively low loading of approximately .40 in each. In the factor of emotional reactivity, this loading is seventh in order of size, and in the factor of field exploration it is fourth. We can conclude that two kinds of defecation occur in this situation, possibly with different adaptive functions. Furthermore, we must conclude that the defecation rate is not the best way to measure even the fear reaction, which is the basis of the construct of emotional reactivity.

Finally, we can conclude from these data that there are at least two kinds of emotional reactivity in the rat and probably many more. This agrees with Fuller's results with emotional reactivity in the dog.[6] These tests gave beautifully clear breed differences and, in the F_1 generation, in which there is little genetic variation, a factorial analysis indicated that emotional reactivity is a single unit. However, in the F_2 generation, in which genetic factors were segregating, emotional reactivity was split between at least two factors, one of which was concerned with heart-rate responses and the other with fear of strange apparatus, and hence lack of success in certain problem-solving situations. Other results indicated that there were several different kinds of timidity rather than any single general trait of fearfulness.

The Development of Social Behavior in the Rat

One obvious feature of some of the experimental designs described by Dr. Denenberg is that the age from birth until 21 days is arbitrarily treated as a unit. In previous experiments, Denenberg and other workers used a 20-day period, arbitrarily divided into 5- or 10-day units. There is no basis for selecting this period, except that it is customary in rat laboratories to wean the pups from their mothers at 21 days, experience having shown that young rats will survive reasonably well after this time. Until recently we have not had the information to permit us to dissect early experience in the rat on a more theoretical basis, but the observational study of Bolles and Woods[1] largely remedies this deficiency.

Before discussing the rat, let me briefly describe development in the dog. Like a good many other mammals, there has been an evolutionary tendency toward a course of development in which the newborn animal is at first behaviorally adapted to an almost completely dependent existence, and later in development shows behavior adapted for a more independent adult existence. Between these two there is a period of very rapid transition, which is almost as spectacular as the metamorphosis from the tadpole to the frog.

As comparative data accumulate, we are beginning to realize that the dog may be unique in that the majority of the transition processes are concentrated in one short period. A dog has a neonatal period of approximately two weeks, a one-week transition period, and then a period of socialization extending from approximately three to twelve weeks, with its peak between six and eight weeks.

Comparison shows that the rat is somewhat different. In the first place, the differentiation between neonatal existence and adult existence is less complete, for adult rats also spend a great deal of time in a nest environment similar to that in which they were born. Second, the newborn rat is even more immature than the dog, as it is born hairless. The rat has a typical neonatal period, with all behavior concentrated around suckling, et-epimeletic (care-soliciting) behavior associated with pain and discomfort, and eliminative behavior elicited by the mother's licking. Transitions begin at ten days of age, with the change from the neonatal form of locomotion to the adult one of walking. Sensory transitions occur at 13 days, when hearing begins, and at 17 days, when vision develops as the eyes open (although this may begin as early as 14 days). The order of development of these major senses is reversed from that in the dog. Transition to the adult mode of nutrition begins on Day 17, correlated with the emergence of vision, and the same day marks transitions to adult patterns of social behavior, particularly social grooming and playful fighting. We can conclude that there is a transition period in the rat from roughly 10 until 17 days, and we would predict that the socialization period would begin after this point, although there is no experimental evidence.

Bolles and Woods weaned their rats at the traditional age of 21 days, but it is obvious from their data that the young rats had not stopped nursing, and since the investigators saw the first fecal boli immediately after weaning, we can conclude that the mothers were still cleaning their young. This means that rats weaned at 21 days, although they survive quite well, are actually being weaned early, which may result in some effect on their emotional development.

Considered from the perspective of development, Denenberg's rats (like the great majority of experimental rats) were not only early weaned, but suddenly weaned. Furthermore, the period of 0–21 days, which he has arbitrarily designated that of early experience, is one in which the young rats go through two distinct developmental periods and part of a third.

CRITICAL PERIODS

The analysis of the critical period phenomenon has enormous practical implications, but also has considerable theoretical interest. The verification of a critical period should proceed as follows. The first step is the empirical discovery of such a phenomenon through observation or experimental manipulation. The results are so obvious and striking that they are unmistakable. Anyone who has seen a critical period effect in the course of his own work has no doubt about its reality.

The second step is the analysis of its basis, which may vary greatly from species to species, and also from one process to another. The basis of a critical period is an organizational process proceeding more rapidly at one time of life than another. The general theoretical principle is that it is easy to modify the results of an organizational process at the time it is proceeding most rapidly, and that it is difficult to modify such a process either after it has stopped and organization is complete, or before it has started. Consequently, the existence of a critical period will depend upon whether an organizational process proceeds at a uniform rate, or proceeds more rapidly at one period than another. It follows that not every organizational process will show a critical period phenomenon, and also that it is difficult to analyze critical period effects experimentally without some knowledge of the timing of the underlying process.

I think this will resolve most of the difficulties which Denenberg raised in his earlier papers[2] concerning the critical period phenomenon, but I would like to raise a question. What is the underlying process that is modified by early stimulation? The work of Levine[4] indicates that this process is the organization of part of the neuroendocrine system, that related to the adrenocortical stress mechanism. It would seem to me that we need some direct anatomical and physiological developmental studies of this mechanism. Once we know the rate and nature of its organization, we can define the existence of a critical period very exactly, if one exists.

EXPERIMENTAL DESIGN AND EARLY EXPERIENCE

From the evidence he has presented here, Denenberg is obviously a master of experimental design, and I have no criticism of his work. Rather, I wish to present a few basic criteria for a good experiment in the special area of early experience.

The experiment should be based on a thorough knowledge of normal behavioral development of the species. One species varies a great deal from another, and the same experiment at birth or hatching or any such arbitrary time will obviously give different results.

The experiment should demonstrate an initial effect on the young organism. The nature of the effect will obviously depend on the state of development. One would not expect visual experience to have much effect on an animal whose eyes have not yet opened.

The experiment should demonstrate an important and lasting effect upon the behavior of the adult animals. Almost any experience will affect the later behavior of an adult animal, but the effects are usually brief and disappear rapidly. The mere demonstration that early experience has an effect on later behavior means very little.

There should be some sort of reasonable theoretical explanation connecting the two events. Raw empirical results are not convincing in themselves, although they may be completely valid.

One measure we found extremely useful in our studies of early experience and behavioral development in the dog is that of physical growth as measured by weighing. This is one of the best measurements of consummatory behavior, and we have found that eating is one of the first things affected by emotional disturbance. The weight of a puppy will be grossly affected within 24 hours by emotional or physiological disturbance. Most of the experiments with rats have used this measure only in a limited way, and I would recommend it wherever the limitations of the experiment permit.

GENERAL COMMENTS

Finally, Denenberg's discovery of effects that extend from one generation to another by nongenetic means has very important implications, both practically in the management of human pregnancies and theoretically for those experimenters who are studying genetic and biological differences. The early experience of the mother affects that of her offspring, and is a variable that must be controlled in behavior genetics experiments with mammals. I also urge the development and use of more good inbred genetic strains of rats whose behavioral characteristics are known. These would not only be useful in identifying genetic components in behavioral variation, but also would reduce the risk of generalizing from a very narrow genetic base.

These nongenetic effects may explain the results of earlier workers who studied the inheritance of wildness and tameness in rats. They found that wild rats raised in the laboratory lost much of their wildness after two or three generations, although there had been no known selection toward that effect.

Regarding human applications of this finding, Pasamanick and Knobloch[5] have found a correlation between disturbances of pregnancy (particularly the toxemias that affect oxygen in the last months of pregnancy) and later behavioral deficiencies of the children. These are much more common among poverty-stricken mothers than others, and would imply a circular effect. A poor prenatal environment tends to produce a handicapped child, who in turn is unable to provide a good prenatal environment for her own children.

In summary, Denenberg's work has been concerned with an empirical study of the organization of emotional behavior, and he has discovered two general factors, which can be labeled fearful behavior and exploratory behavior. The further conclusion can be drawn that both kinds of behavioral organization draw on some of the same emotional components. In the rat, at least, there does not appear to be a completely separable emotional mechanism for each of the behavioral systems associated with general function. Rather, a limited number of emotional mechanisms are drawn upon in different combinations and in different degrees. Only one part of the nervous system acts in a unitary fashion in connection with emotion — the sympathetic system in connection with anger. Even here it has been shown that the reactions are largely of a physiological preparatory nature and have little effect on behavior except to modify its efficiency. In sexual behavior, elements of both the sympathetic and parasympathetic systems are involved, and so on. We would further predict that the organization of behavior and its emotional bases would differ from one species to another and very likely from one individual to another within the same species. This again emphasizes the importance of using known genetic material.

Analysis of Infant Stimulation

JOHN W. M. WHITING

I would like to commend Professor Denenberg on the research program that he and his colleagues have been carrying out. The studies reported are carefully designed, cumulative in effect, and carry us a long step forward in our knowledge of the effects of infant stimulation.

Denenberg's analysis of the consequences of infant stimulation is particularly valuable. Some of the apparently contradictory results of previous studies have been cleared up by his factor analytic procedures. Furthermore, his demonstration that animals stressed in infancy are apparently more curious and prefer variation in both tactile and visual stimulation represents a new and interesting finding.

His attempts at manipulating the independent variable by controlling the emotionality of the mother are less successful. Not only are the results contradictory, but some of the experiments showed no first-order effects at all. In two of the studies reported,[2,7] emotional mothers reared pups that were *more* emotional than the controls when tested at adulthood; in a third,[3] the pups were *less* emotional. Finally, in a fourth experiment,[1] "the characteristics of the mother were not found to have any significant main effect upon the several measures of emotionality, although this variable was involved in a few significant interactions." On the other hand, the standard procedure of "handling" produced consistent results. These findings suggest that the "emotional" mothers do not behave toward their litters in any consistent way that is more or less stimulating to the pups and that one is, in a sense, "trusting to luck" by using the mother's emotionality as an independent variable.

On the other hand, if mother rats were carefully observed in their mothering techniques, and could be distinguished on some variables such as whether or not they interrupted nursing, avoided the pup-

---

JOHN W. M. WHITING Harvard University

pies when not nursing, or were more or less rough with them, more consistent results might be expected. I think perhaps Denenberg was led astray in this part of his research by a bad strategy. He says, "Our interest here has been in trying to set up analogs to investigate hypotheses derived from human observations in a clinical context." If so, we should know much more precisely what kinds of human mothering behavior produce effects of the kind described by Denenberg. Most clinical theories involve distrust, identification, or rejection — theories that seem inappropriate for accounting for such effects.

Before using the emotionality of the mother as a variable, I think a much more careful investigation of the nature and intensity of infant stimulation should be carried out. The standard procedure in animal experiments is, as Denenberg states, "to remove the pups from the home cage . . . place them in a container where they remain for a brief period of time . . . and then return them to the home cage." The effect of this treatment has been variously interpreted: A) it gets the animal used to the experimenter; B) it is stimulating in the sense of stimulus variation, or enrichment; and C) it is stressful or frightening. That all the other procedures found to be effective — electric shock, vibration, cold — seem to be relatively painful, favors the interpretation that "handling" procedures are stressful. Research on primates by Harlow, DeVore, and others indicates that rhesus and baboon infants are easily frightened when not in body contact with the mother. This suggests that even the apparently mild stimulation involved in human handling procedures may indeed be frightening to the pup.

This stress interpretation seems the most plausible way to account for the human material, also. The customary treatment of infants reported by Landauer and Whiting[6] to be associated with an increase in stature includes piercing the nose, lips or ears, circumcision, inoculation, scarification, and cauterization, all of which seem to be variations in the intensity rather than the diversity of stimulation. Daily stretching of the limbs or molding the head are procedures less clearly interpretable, but it is not impossible that such manipulations could evoke a stress response in a very young infant.

Systematic separation of human infants from the mother has also been shown to increase growth.[4] In a cross-cultural study similar in design to that of Landauer and Whiting, Gunders found that adult males were taller on the average in societies where infants were not nursed for the first 12 hours, were ritually held or handled by someone other than the mother, and not continuously held or carried in body contact with the mother. These procedures parallel the "handling" procedures for rats. In a study of Yemenite children in Israel,[5] Dr. Gunders and I found that those born in hospitals and separated from the mother in accordance with standard hospital procedures were significantly heavier at two, three, and four years of age than were those children born at home, where they were in much more constant contact with the mother.

Even though it may seem plausible that the various treatments described above have in common the effect of stressing or frightening the infant, there has not been, as far as I am aware, any successful attempt to measure directly the immediate physiological effects of infant stimulation. Of course, great technical problems are involved in such an attempt, but I do not think they are insurmountable. Until this is done, I am afraid that infant stimulation and its effects will remain somewhat mysterious.

I think another line of research is even more badly needed. This is suggested by Denenberg in his first hypothesis — that emotional reactivity is reduced as a monotonic function of the amount of stimulus input in infancy. He does not, however, present any empirical data to support the assumption that the function is monotonic. There is even less empirical justification that the function is of the shape shown in his Figure 1. I suspect that the function is more likely to be curvilinear. Very strong stimulation may have less effect in reducing emotional reactivity in adulthood than does mild or moderate stimulation. It may indeed have the effect of increasing reactivity. But my suspicion is based on no more evidence than Denenberg's. Research is badly needed in which the intensity of stimulation is systematically varied.

References

KARL H. PRIBRAM. *Steps Toward a Neuropsychological Theory*

1 ADEY, W. R., R. T. KADO, AND J. DIDIO. Impedance measurements in brain tissue of animals using microvolt signals, *Exptl. Neurol.*, 1962, Vol. 5, pp. 47–66.
2 ANDERSEN, P. AND J. C. ECCLES. Inhibitory phasing of neuronal discharge, *Nature*, 1962, Vol. 196, pp. 645–647.
3 ANDERSEN, P., J. C. ECCLES, AND T. A. SEARS. Presynaptic inhibitory action of cerebral cortex on the spinal cord, *Nature*, 1962, Vol. 194, pp. 740–741.
4 ARNOLD, M. B. Emotion and Personality, Vol. II: Neurological and Physiological Aspects, New York, Columbia Univ. Press, 1960.
5 ASANUMA, H. AND V. B. BROOKS. Recurrent cortical effects following stimulation of internal capsule, *Arch. Ital. Biol.*, 1965, Vol. 103, pp. 220–246.
6 BAGSHAW, M. H., D. P. KIMBLE, AND K. H. PRIBRAM. The GSR of monkeys during orienting and habituation and after ablation of the amygdala, hippocampus and inferotemporal cortex, *Neuropsychologia*, 1965, Vol. 3, pp. 111–119.
7 BARD, P. AND D. McK. RIOCH. A study of four cats deprived of neocortex and additional portions of the forebrain, *Bull. Johns Hopkins Hosp.*, 1937, Vol. 60, pp. 73–147.
8 BARRATT, E. S. Relationship of psychomotor tests and EEG variables at three developmental levels, *Perceptual and Motor Skills*, 1959, Vol. 9, pp. 63–66.
9 BARRATT, E. S. Anxiety and impulsiveness related to psychomotor efficiency, *Perceptual and Motor Skills*, 1959, Vol. 9, pp. 191–198.
10 BATESON, P. (Unpublished data)
11 BECHTEREV, W. VON. Die Funktionen Der Nervencentral, Berlin, Fisher Verlag, 1911.
12 BÉKÉSY, G. V. Neural volleys and the similarity between some sensations produced by tones and by skin vibrations, *J. Acoust. Soc. Am.*, 1957, Vol. 29, pp. 1059–1069.
13 BROOKS, V. B. AND H. ASANUMA. Recurrent corticial effects following stimulation of medullary pyramid, *Arch. Ital. Biol.*, 1965, Vol. 103, pp. 247–278.
14 BROOKS, V. B. AND H. ASANUMA. Pharmacological studies of recurrent cortical inhibition and facilitation, *Am. J. Physiol.*, 1965, Vol. 208, pp. 674–681.
15 CANNON, W. B. The James-Lange theory of emotions: a critical examination and an alternative theory, *Am. J. Psychol.*, 1927, Vol. 39, pp. 106–124.
16 DELGADO, J. M. R. Brain centers and control of behavior–animals, *in* The First Hahnemann Symposium on Psychosomatic Medicine, 1962, pp. 221–227.

17 DELGADO, R. R. AND J. M. R. DELGADO. An objective approach to measurement of behavior, *Phil. Sci.*, 1962, Vol. 29, pp. 253–268.
18 DEMENT, W. C. An essay on dreams: the role of physiology in understanding their nature, *in* New Directions in Psychology, Vol. II, New York, Holt, Rinehart and Winston, 1965, pp. 137–257.
19 DEWSON III, J. H., K. NOBEL, AND K. H. PRIBRAM. Corticofugal influence at cochlear nucleus of the cat. Accepted for publication, *J. Acoust. Soc. Amer.*, Nov. 1965.
20 DOUGLAS, R. J. AND R. L. ISAACSON. Hippocampal lesions and activity, *Psychonomic Sci.*, 1964, Vol. 1, pp. 187–188.
21 DOUGLAS, R. AND K. H. PRIBRAM. Learning and limbic lesions, *Neuropsychologia*, 1966, Vol. 4, pp. 197–220.
22 ECCLES, J. C. Inhibitory controls on the flow of sensory information in the nervous system, *in* Information Processing in the Nervous System, Vol. III, Proceedings of the International Union of Physiological Sciences, XXII International Congress of Physiological Sciences, Leiden, 1962, pp. 24–48.
23 FAIR, C. M. The Physical Foundations of the Psyche, Middletown, Wesleyan Univ. Press, 1963.
24 Fox, S. S. *Progr. Brain Res.*, 1966, Vol. 27 (in press).
25 GROSSMAN, S. P. The VMH: a center for affective reactions, satiety, or both? *in* Physiology and Behavior, 1966, Vol. I, Pergamon Press, pp. 1–10.
26 HARTLINE, H. K., H. G. WAGNER, AND F. RATLIFF. Inhibition in the eye of *Limulus*, *J. Gen. Physiol.*, 1956, Vol. 39, pp. 651–673.
27 HEAD, H. Studies in Neurology, London, M. Frowde; Hodder and Stoughton, 1920.
28 HESS, W. R. Diencephalon: Autonomic and Extrapyramidal Functions, New York, Grune and Stratton, 1954.
29 KAADA, B. R., K. H. PRIBRAM, AND J. A. EPSTEIN. Respiratory and vascular responses in monkeys from temporal pole, insula, orbital surface and cingulate gyrus. A preliminary report, *J. Neurophysiol.*, 1949, Vol. 12, pp. 347–356.
30 KETY, S. S. Catecholamines in neuropsychiatric states, *Pharmacol. Rev.*, 1966, Vol. 18, pp. 787–798.
31 KIMBLE, D. P. The effects of bilateral hippocampal lesions in rats, *J. Comp. Physiol. Psychol.*, 1963, Vol. 56, pp. 273–283.
32 KIMBLE, D. P., M. BAGSHAW, AND K. H. PRIBRAM. The GSR of monkeys during orienting and habituation after selective partial ablations of the cingulate and frontal cortex, *Neuropsychologia*, 1965, Vol. 3, pp. 121–128.
33 KLEITMAN, N. Sleep and Wakefulness, Chicago, Univ. of Chicago Press, 1963.
34 KOEPKE, J. E. AND K. H. PRIBRAM. Habituation of GSR as a function of stimulus duration and spontaneous activity, *J. Comp. Physiol. Psychol.*, 1966, Vol. 61, pp. 442–448.
35 KRASNE, F. B. General disruption resulting from electrical stimulus of ventromedial hypothalamus, *Science*, 1962, Vol. 138, pp. 822–823.
36 LACEY, J. I., J. KAGAN, B. C. LACEY, AND H. A. MOSS. The visceral level: situational determinants and behavioral correlates of autonomic response patterns,

in Expressions of the Emotions in Man (P. H. Knapp, editor), New York, International Universities Press, 1963, pp. 161–208.

37 Lashley, K. The thalamus and emotion, *in* The Neuropsychology of Lashley (F. A. Beach, D. O. Hebb, C. T. Morgan, and H. W. Nissen, editors), New York, McGraw-Hill, 1960, pp. 345–360.

38 Li, C.-L., C. Cullen, and H. H. Jasper. Laminar microelectrode analysis of cortical unspecific recruiting responses and spontaneous rhythms, *J. Neurophysiol.*, 1956, Vol. 19, pp. 131–143.

39 Li, C.-L., C. Cullen, and H. H. Jasper. Laminar microelectrode studies of specific somatosensory cortical potentials, *J. Neurophysiol.*, 1956, Vol. 19, pp. 111–130.

40 Lindsley, D. B. Emotion, *in* Handbook of Experimental Psychology (S. S. Stevens, editor), New York, Wiley, 1951, pp. 473–516.

41 MacLean, P. D. Psychosomatic disease and the "visceral brain," recent developments bearing on the Papez theory of emotion, *Psychosomat. Med.*, 1949, Vol. 11, pp. 338–353.

42 Mandler, G. The interruption of behavior, *in* Nebraska Symposium on Motivation (D. Levine, editor), Lincoln, Univ. of Nebraska Press, 1964, pp. 163–220.

43 McCleary, R. A. Response specificity in the behavioral effects of limbic system lesions in the cat, *J. Comp. Physiol. Psychol.*, 1961, Vol. 54, pp. 605–613.

44 Miller, G. A., E. Galanter, and K. H. Pribram. Plans and the Structure of Behavior, New York, Henry Holt, 1960.

45 Miller, N. E., C. J. Bailey, and J. A. F. Stevenson. Decreased "hunger" but increased food intake resulting from hypothalamic lesions, *Science*, 1950, Vol. 112, pp. 256–259.

46 Milner, B. Psychological defects produced by temporal lobe excision, *Res. Publ. Assoc. Res. Nervous Mental Disease*, 1958, Vol. 36, pp. 244–257.

47 Mountcastle, V. B. Modality and topographic properties of single neurons of cat's sensory cortex, *J. Neurophysiol.*, 1957, Vol. 20, pp. 408–434.

48 Papez, J. W. A proposed mechanism of emotion, *Arch. Neurol. Psychiat.*, 1937, Vol. 38, pp. 725–743.

49 Penfield, W. and B. Milner. Memory deficit produced by bilateral lesions in the hippocampal zone, A.M.A. *Arch. Neurol. Psychiat.*, 1958, Vol. 79, pp. 475–497.

50 Peters, R. S. Emotions, passivity, and the place of Freud's theory in psychology, *in* Scientific Psychology: Principles and Approaches (B. B. Wolman and E. Nagel, editors), New York, Basic Books, 1965, pp. 365–383.

51 Pribram, K. H. Toward a science of neuropsychology: (method and data), *in* Current Trends in Psychology and the Behavioral Sciences (R. A. Patton, editor), Pittsburgh, Univ. of Pittsburgh Press, 1954, pp. 115–142.

52 Pribram, K. H. Comparative neurology and the evolution of behavior, *in* Behavior and Evolution (Anne Roe and G. G. Simpson, editors), New Haven, Yale Univ. Press, 1958, pp. 140–164.

53 Pribram, K. H. A review of theory in physiological psychology, *Ann. Rev. Psychol.*, 1960, Vol. 11, pp. 1–40.

54 PRIBRAM, K. H. Reinforcement revisited: a structural view, *in* Nebraska Symposium on Motivation (M. R. Jones, editor), Lincoln, Univ. of Nebraska Press, 1963, pp. 113–159.
55 PRIBRAM, K. H. Memory and the organization of attention and intention: the case history of a model, *in* Brain Function and Learning (V. P. Hall, editor), Los Angeles, Univ. of California Press (in press).
56 PRIBRAM, K. H. The Limbic Systems, Efferent Control of Neural Inhibition and Behavior, Prog. Brain Res., (T. Tokizane and J. P. Schade, editors) 1966 (in press).
57 PRIBRAM, K. H. A neuropsychological analysis of cerebral function: an informal progress report of an experimental program, *Canadian Psychologist*, 1966, Vol. 7a, Inst. Suppl., pp. 324–367.
58 PRIBRAM, K. H., K. W. GARDNER, G. L. PRESSMAN, AND M. BAGSHAW. An automated discrimination apparatus for discrete trial analysis (DADTA), *Psychol. Rept.*, 1962, Vol. 11, pp. 247–250.
59 PRIBRAM, K. H. AND L. KRUGER. Functions of the "olfactory brain," *Ann. N.Y. Acad. Sci.*, 1954, Vol. 58, pp. 109–138.
60 PRIBRAM, K. H. AND F. T. MELGES. Emotion: the search for control, *in* Handbook of Clinical Neurology (P. Vinken and G. Bruyn, editors), Amsterdam, North-Holland (in press).
61 ROBERTS, W. W., W. N. DEMBER, AND M. BRODWICK. Alternation and exploration in rats with hippocampal lesions, *J. Comp. Physiol. Psychol.*, 1962, Vol. 55, pp. 695–700.
62 ROTHSTEIN, D. A. Psychiatric implications of information theory, *Arch. Gen. Psychiat.*, 1965, Vol. 13, pp. 87–94.
63 SCHACHTER, S. This volume.
64 SCHWARTZBAUM, J. S. Changes in reinforcing properties of stimuli following ablation of the amygdaloid complex in monkeys, *J. Comp. Physiol. Psychol.*, 1960, Vol. 53, pp. 388–395.
65 SCHWARTZBAUM, J. S. Response to changes in reinforcing conditions of bar-pressing after ablation of the amygdaloid complex in monkeys, *Psychol. Rept.*, 1960, Vol. 6, pp. 215–221.
66 SCHWARTZBAUM, J. S. Some characteristics of amygdaloid hyperphagia in monkeys, *Am. J. Psychol.*, 1961, Vol. 74, pp. 252–259.
67 SCHWARTZBAUM, J. S. Visually reinforced behavior following ablation of the amygdaloid complex in monkeys, *J. Comp. Physiol. Psychol.*, 1964, Vol. 57, pp. 340–347.
68 SOKOLOV, E. H. Neuronal models and the orienting reflex, *in* The Central Nervous System and Behavior (M. A. B. Brazier, editor), New York, Josiah Macy Jr. Foundation, 1960, pp. 187–276.
69 SPENCER, W. A., R. F. THOMPSON, AND D. R. NEILSON, JR. Response decrement of the flexion reflex in the acute spinal cat and transient restoration by strong stimuli, *J. Neurophysiol.*, 1966, Vol. 29, pp. 221–239.
70 SPENCER, W. A., R. F. THOMPSON, AND D. R. NEILSON, JR. Alterations in responsiveness of ascending and reflex pathways activated by iterated cutaneous afferent volleys, *J. Neurophysiol.*, 1966, Vol. 29, pp. 240–252.

71 SPENCER, W. A., R. F. THOMPSON, AND D. R. NEILSON, JR. Decrement of ventral root electrotonus and intracellularly recorded PSPs produced by iterated cutaneous afferent volleys, *J. Neurophysiol.*, 1966, Vol. 29, pp. 253–274.

72 SPINELLI, D. N. AND K. H. PRIBRAM. Changes in visual recovery functions produced by temporal lobe stimulation in monkeys, *Electroencephalog. Clin. Neurophysiol.*, 1966, Vol. 20, pp. 44–49.

73 SPINELLI, D. N. AND K. H. PRIBRAM. Changes in visual recovery functions and unit activity produced by frontal and temporal cortex stimulation, *Electroencephalog. Clin. Neurophysiol.*, 1967, Vol. 22, pp. 143–149.

74 STELLAR, E. The physiology of motivation, *Psychol. Rev.*, 1954, Vol. 61, pp. 5–22.

75 WALL, P. D. AND K. H. PRIBRAM. Trigeminal neurotomy and blood pressure responses from stimulation of lateral cerebral cortex of *Macaca mulatta*, *J. Neurophysiol.*, 1950, Vol. 13, pp. 409–412.

76 WATERMAN, T. H. Systems analysis and the visual orientation of animals, *Am. Scientist*, 1966, Vol. 54, pp. 15–45.

ACKNOWLEDGMENT

This research was supported by NIMH Career Award MH–15,214, NIMH Grant MH–03732, and Dept. of the Army Contract MD–2328. I am especially indebted to Dr. Fred Melges who collaborated with me on an extension of this proposal, and to Mr. Walter Tubbs and Mrs. Phyllis Ellis, who retyped and edited the many revisions.

JOHN P. FLYNN. *The Neural Basis of Aggression in Cats*

1 ADAMS, D. AND J. P. FLYNN. Transfer of an escape response from tail shock to brain stimulated attack behavior, *J. Exp. Anal. Behav.*, 1966, Vol. 9, pp. 401–408.

2 BARD, P. A diencephalic mechanism for the expression of rage with special reference to the sympathetic nervous system, *Amer. J. Physiol.*, 1928, Vol. 84, pp. 490–515.

3 BARD, P. AND V. B. MOUNTCASTLE. Some forebrain mechanisms involved in expression of rage with special reference to suppression of angry behaviour, *Res. Publ. Ass. Res. Nervous Mental Disease*, 1948, Vol. 27, pp. 362–404.

4 BARD, P. AND D. McK. RIOCH. A study of four cats deprived of neocortex and additional portions of the forebrain, *Bull. Johns Hopkins Hosp.*, 1937, Vol. 60, pp. 73–147.

5 DELGADO, J. M. R., W. W. ROBERTS, AND N. E. MILLER. Learning motivated by electrical stimulation of the brain, *Am. J. Physiol.*, 1954, Vol. 179, pp. 587–593.

6 EGGER, M. D. AND J. P. FLYNN. Effects of electrical stimulation of the amygdala on hypothalamically elicited attack behavior in cats, *J. Neurophysiol.*, 1963, Vol. 26, No. 5, pp. 705–720.

7 ELLISON, G. D. AND J. P. FLYNN. Aggressive behavior in cats after surgical isolation of the hypothalamus. (In preparation)
8 GOLTZ, F. Der Hund ohne Grosshirn, *in* Siebente Abhandlung über die Verrichtungen des Grosshirns, *Pflügers Arch. ges. Physiol.*, 1892, Vol. 51, pp. 570–614.
9 HESS, W. R. Stammganglien-Reizversuche, 10. Tagung der Deutschen Physiologischen Gesellschaft, Frankfurt am Main, *Ber. ges. Physiol*, 1928, Vol. 42, pp. 554–555.
10 HESS, W. R. AND K. AKERT. Experimental data on role of hypothalamus in mechanism of emotional behavior, *Arch. Neurol. Psychiat.*, 1955, Vol. 73, No. 2, pp. 127–129.
11 HESS, W. R. AND M. BRÜGGER. Das subkortikale Zentrum der affektiven Abwehrreaktion, *Helv. Physiol. Pharmacol. Acta*, 1943, Vol. 1, pp. 35–52.
12 HUNSPERGER, R. W. Affektreaktionen auf elektrische Reizung im Hirnstamm der Katze, *Helv. Physiol. Pharmacol. Acta*, 1956, Vol. 14, pp. 70–92.
13 HUTCHINSON, R. R. AND J. W. RENFREW. Stalking attack and eating behaviors elicited from the same sites in the hypothalamus, *J. Comp. Physiol. Psychol.*, 1966, Vol. 61, No. 3, pp. 360–367.
14 KLÜVER, H. AND P. C. BUCY. Preliminary analysis of functions of the temporal lobes in monkeys, *Arch. Neurol. Psychiat.*, 1939, Vol. 42, pp. 979–1000.
15 LEVISON, P. K. AND J. P. FLYNN. The objects attacked by cats during stimulation of the hypothalamus, *Anim. Behav.*, 1965, Vol. 13, pp. 217–220.
16 LINDSLEY, D. B. Emotion, *in* Handbook of Experimental Psychology (S. S. Stevens, editor), New York, Wiley, 1951, pp. 473–516.
17 MACDONNELL, M. F. AND J. P. FLYNN. Attack elicited by stimulation of the thalamus of cats, *Science*, 1964, Vol. 144, pp. 1249–1250.
18 MACDONNELL, M. F. AND J. P. FLYNN. Control of sensory fields by stimulation of hypothalamus, *Science*, 1966, Vol. 152, pp. 1406–1408.
19 MACDONNELL, M. F. AND J. P. FLYNN. Sensory control of hypothalamic attack, *Anim. Behav.* 1966, Vol. 14, pp. 339–405.
20 MACLEAN, P. D. Psychosomatic disease and the "visceral brain." Recent developments bearing on the Papez theory of emotion, *Psychosomat. Med.*, 1949, Vol. 11, No. 6, pp. 338–353.
21 MASSERMAN, J. H. Is the hypothalamus a center of emotion? *Psychosomat. Med.*, 1941, Vol. 3, No. 1, pp. 3–25.
22 MORUZZI, G. AND H. W. MAGOUN. Brain stem reticular formation and activation of the EEG, *Electroenceph. Clin. Neurophysiol.*, 1949, Vol. 1, No. 4, pp. 455–473.
23 NAKAO, H. Emotional behavior produced by hypothalamic stimulation, *Am. J. Physiol.*, 1958, Vol. 194, pp. 411–418.
24 PAPEZ, J. W. A proposed mechanism of emotion, *Arch. Neurol. Psychiat.*, 1937, Vol. 38, pp. 725–743.
25 RANSON, S. W. The hypothalamus: its significance for visceral innervation and emotional expression. The Weir Mitchell Oration, *Trans. Coll. Physns. Philad.* Series IV, 1934, Vol. 2, No. 3, pp. 222–242.

26 ROTHMANN, H. Zusammenfassender Bericht über den Rothmannschen grosshirnlosen Hund nach klinischer und anatomischer Untersuchung, *Ges. Neurol. Psychiat.*, Berl., 1923, Vol. 87, pp. 247–313.
27 SCHALTENBRAND, G. AND S. COBB. Clinical and anatomical studies on two cats without neocortex, *Brain*, 1930–31, Vol. 53, pp. 449–488.
28 SHEARD, M. AND J. P. FLYNN. Facilitation of attack behavior by stimulation of the midbrain of cats, *Brain Research*, 1967, Vol. 4, pp. 324–333.
29 SCHREINER, L. AND A. KLING. Behavioral changes following rhinencephalic injury in cat, *J. Neurophysiol.*, 1953, Vol. 16, No. 6, pp. 643–659.
30 SIEGEL, A. AND J. P. FLYNN. Differential effects of hippocampal stimulation upon attack elicited by stimulation of the hypothalamus. (In preparation)
31 WASMAN, M. AND J. P. FLYNN. Directed attack elicited from hypothalamus, *Arch. Neurol.*, 1962, Vol. 6, pp. 220–227.
32 WASMAN, M. AND J. P. FLYNN. Directed attack behavior during hippocampal seizures, *Arch. Neurol.*, 1966, Vol. 14, pp. 408–414.
33 WOODWORTH, R. S. AND C. S. SHERRINGTON. A pseudaffective reflex and its spinal path, *J. Physiol.*, 1904, Vol. 31, pp. 234–243.

RONALD MELZACK. *Brain Mechanisms and Emotion: Discussion of Karl Pribram's Paper*

1 BORING, E. G. A History of Experimental Psychology (2nd ed.), New York, Appleton-Century-Crofts, 1950.
2 DALLENBACH, K. M. Pain: history and present status, *Am. J. Psychol.*, 1939, Vol. 52, No. 3, pp. 331–347.
3 DELGADO, J. M. R., W. W. ROBERTS, AND N. E. MILLER. Learning motivated by electrical stimulation of the brain, *Am. J. Physiol.*, 1954, Vol. 179, pp. 587–593.
4 FORD, F. R. AND L. WILKINS. Congenital universal insensitiveness to pain, *Bull. Johns Hopkins Hosp.*, 1938, Vol. 62, pp. 448–466.
5 FREEMAN, W. AND J. W. WATTS. Pain mechanisms and the frontal lobes: a study of prefrontal lobotomy for intractable pain, *Ann. Internal Med.*, 1948, Vol. 28, No. 4, pp. 747–754.
6 GRASTYÁN, E., J. CZOPF, L. ÁNGYÁN, AND I. SZABÓ. The significance of subcortical motivational mechanisms in the organization of conditional connections, *Acta Physiol. Acad. Sc. Hung.*, 1965, Vol. 26, pp. 9–46.
7 HEBB, D. O. On the nature of fear, *Psychol. Rev.*, 1946, Vol. 53, No. 5, pp. 259–276.
8 HEBB, D. O. Drives and the C.N.S. (conceptual nervous system), *Psychol. Rev.*, 1955, Vol. 62, pp. 243–254.
9 HEBB, D. O. A Textbook of Psychology (2nd ed.), Philadelphia, W. B. Saunders, 1966.
10 KAPLAN, M. The effects of noxious stimulus intensity and duration during intermittent reinforcement of escape behavior, *J. Comp. Physiol. Psychol.*, 1952, Vol. 45, pp. 538–549.

11 KING, H. E., J. CLAUSEN, AND J. E. SCARFF. Cutaneous thresholds for pain before and after unilateral prefrontal lobotomy, *J. Nervous Mental Disease,* 1950, Vol. 112, No. 2, pp. 93–96.
12 LINDSLEY, D. B. Emotion, *in* Handbook of Experimental Psychology (S. S. Stevens, editor), New York, Wiley, 1951, pp. 473–516.
13 MARSHALL, H. R. Pain, Pleasure, and Aesthetics, London, Macmillan, 1894.
14 MELZACK, R. Irrational fears in the dog, *Can. J. Psychol.,* 1952, Vol. 6, pp. 141–147.
15 MELZACK, R. The perception of pain, *Sci. Am.,* 1961, Vol. 204, No. 2, pp. 41–49.
16 MELZACK, R. Effects of early experience on behavior: experimental and conceptual considerations, *in* Psychopathology of Perception (P. H. Hoch and J. Zubin, editors), New York, Grune and Stratton, 1965, pp. 271–299.
17 MELZACK, R. AND K. L. CASEY. The sensory, motivational, and central control determinants of pain, *in* International Symposium on the Skin Senses (D. Kenshalo, editor), Springfield, C. C Thomas, 1967 (in press).
18 MELZACK, R. AND P. D. WALL. Pain mechanisms: a new theory, *Science,* 1965, Vol. 150, pp. 971–979.
19 OLDS, J. Physiological mechanisms of reward, *in* Nebraska Symposium on Motivation (M. R. Jones, editor), Lincoln, Univ. of Nebraska Press, 1955, pp. 73–139.
20 PFAFFMANN, C. Behavioral responses to taste and odor stimuli, *in* Flavor Research and Food Acceptance, New York, Reinhold, 1958, pp. 29–44.
21 PFAFFMANN, C. The pleasures of sensation, *Psychol. Rev.,* 1960, Vol. 67, No. 4, pp. 253–268.
22 SCHILDER, P. AND E. STENGEL. Asymbolia for pain, *Arch. Neurol. Psychiat.,* 1931, Vol. 25, No. 3, pp. 598–600.
23 SHERRINGTON, C. S. Cutaneous sensations, *in* Textbook of Physiology, Vol. 2, (E. A. Schäfer, editor), Edinburgh, Pentland, 1900, pp. 920–1001.
24 STEWART, J. Reinforcing effects of light as a function of intensity and reinforcement schedule, *J. Comp. Physiol. Psychol.,* 1960, Vol. 53, pp. 187–193.
25 STRONG, C. A. The psychology of pain, *Psychol. Rev.,* 1895, Vol. 2, pp. 329–347.
26 SWEET, W. H. Pain, *in* Handbook of Physiology, Section 1: Neurophysiology, Vol. I, Washington, D.C., American Physiological Soc., 1959, pp. 459–506.
27 YOUNG, P. T. The role of affective processes in learning and motivation, *Psychol. Rev.,* 1959, Vol. 66, pp. 104–125.

JOSEPH V. BRADY. *Emotion and the Sensitivity of Psychoendocrine Systems*

1 BRADY, J. V. Experimental studies of psychophysiological responses to stressful situations, *in* Proc. of Symposium on Medical Aspects of Stress in the Military Climate, Walter Reed Army Institute of Research, U.S. Govt. Printing Off., 1966, pp. 271–295.

2 BRADY, J. V. Psychophysiology of emotional behavior, *in* Experimental Foundations of Clinical Psychology (A. J. Bachrach, editor), New York, Basic Books, 1962, pp. 343–385.
3 CANNON, W. B. Bodily Changes in Pain, Hunger, Fear and Rage, New York, D. Appleton-Century, 1915.
4 ESTES, W. K. AND B. F. SKINNER. Some quantitative properties of anxiety. *J. Exp. Psychol.*, 1941, Vol. 29, No. 5, pp. 390–400.
5 HUNT, H. F. AND J. V. BRADY. Some effects of electro-convulsive shock on a conditioned emotional response ("anxiety"), *J. Comp. Physiol. Psychol.*, 1951, Vol. 44, pp. 88–98.
6 HUNT, H. F., P. JERNBERG, AND J. V. BRADY. The effect of electro-convulsive shock (ECS) on a conditioned emotional response: the effect of post-ECS extinction on the reappearance of the response, *J. Comp. Physiol. Psychol.*, 1952, Vol. 45, pp. 589–599.
7 MASON, J. W. Restraining chair for the experimental study of primates, *J. Appl. Physiol.*, 1958, Vol. 12, pp. 130–133.
8 MASON, J. W. Psychological influences on the pituitary-adrenal cortical system, *in* Recent Progress in Hormone Research (G. Pincus, editor), 1959, Vol. XV, New York, Academic Press, pp. 345–389.
9 MASON, J. W. AND J. V. BRADY. Plasma 17-hydroxycorticosteroid changes related to reserpine effects on emotional behavior, *Science*, 1956, Vol. 124, pp. 983–984.
10 MASON, J. W. AND J. V. BRADY. The sensitivity of psychoendocrine systems to social and physical environment, *in* Psychobiological Approaches to Social Behavior (P. H. Leiderman and D. Shapiro, editors), Stanford, Stanford Univ. Press, 1964, pp. 4–23.
11 MASON, J. W., J. V. BRADY, E. POLISH, J. A. BAUER, J. A. ROBINSON, R. M. ROSE, AND E. D. TAYLOR. Patterns of corticosteroid and pepsinogen change related to emotional stress in the monkey, *Science*, 1961, Vol. 133, pp. 1596–1598.
12 MASON, J. W., J. V. BRADY, AND M. SIDMAN. Plasma 17-hydroxycorticosteroid levels and conditioned behavior in the rhesus monkey, *Endocrinology*, 1957, Vol. 60, No. 6, pp. 741–752.
13 MASON, J. W., J. V. BRADY, AND W. W. TOLSON. Behavioral adaptations and endocrine activity, *in* Endocrines and the Central Nervous System, Proceedings of the Assoc. for Res. Mental Diseases (R. Levine, editor), Vol. 43, Baltimore, Williams and Wilkins, 1966, pp. 227–248.
14 MASON, J. W., J. V. BRADY, W. W. TOLSON, J. A. ROBINSON, E. D. TAYLOR, AND E. H. MOUGEY. Patterns of thyroid, gonadal, and adrenal hormone secretion related to psychological stress in the monkey. *Psychosomat. Med.*, 1961, Vol. 23, No. 5, p. 446 (Abstract).
15 MASON, J. W., G. MANGAN, J. V. BRADY, D. CONRAD, AND D. McK. RIOCH. Concurrent plasma epinephrine, norepinephrine and 17-hydroxycorticosteroid levels during conditioned emotional disturbances in monkeys. *Psychosomat. Med.*, 1961, Vol. 23, No. 4, pp. 344–353.
16 MICHAEL, R. P. AND J. L. GIBBONS. Interrelationships between the endocrine

system and neuropsychiatry, *Intern. Rev. Neurobiol.*, 1963, Vol. 5, pp. 243–302.
17 RAZRAN, G. The observable unconscious and the inferable conscious in current Soviet psychophysiology: interoceptive conditioning, semantic conditioning, and the orienting reflex, *Psychol. Rev.*, 1961, Vol. 68, No. 2, pp. 81–147.
18 SCHUSTER, C. R. AND J. V. BRADY. The discriminative control of a food reinforced operant by interoceptive stimulation, *Pavlov J. of Higher Nervous Activity*, 1964, Vol. 14, No. 3, pp. 448–458.
19 SLUCKI, H., G. ADAM, AND R. W. PORTER. Operant discrimination of an interoceptive stimulus in rhesus monkeys. *J. Exp. Anal. Behav.*, 1965, Vol. 8, No. 6, pp. 405–414.

ACKNOWLEDGMENT

This work was supported in part by National Aeronautics and Space Administration Grant NsG189–61 and Public Health Service Research Grant MH–01604–09 to the University of Maryland.

GEORGE MANDLER. *The Conditions for Emotional Behavior*

1 MANDLER, G. Emotion, *in* New Directions in Psychology (R. W. Brown, E. Galanter, E. H. Hess, and G. Mandler, editors), Vol. I, Holt, Rinehart and Winston, 1962, pp. 267–343.
2 MANDLER, G. The interruption of behavior, *in* Nebraska Symposium on Motivation (D. Levine, editor), Lincoln, Univ. of Nebraska Press, 1964, pp. 163–219.
3 MANDLER, G., JEAN M. MANDLER, I. KREMEN, AND R. D. SHOLITON. The response to threat: Relations among verbal and physiological indices, *Psychol. Monogr.*, 1961, Vol. 75, No. 9, pp. 1–22.
4 MANDLER, G. AND D. L. WATSON. Anxiety and the interruption of behavior, *in* Anxiety and Behavior (C. D. Spielberger, editor), New York, Academic Press, 1966.
5 MASON, J. W., G. MANGAN, JR., J. V. BRADY, D. CONRAD, AND D. McK. RIOCH. Concurrent plasma epinephrine, norepinephrine and 17-hydroxycorticosteroid levels during conditioned emotional disturbances in monkeys, *Psychosomat. Med.*, 1961, Vol. 23, No. 4, pp. 344–353.
6 MOWRER, O. H. AND P. VIEK. An experimental analogue of fear from a sense of helplessness, *J. Abnorm. Soc. Psychol.*, 1948, Vol. 43, No. 2, pp. 193–200.
7 SCHACHTER, S. AND B. LATANÉ. Crime, cognition, and the autonomic nervous system, *in* Nebraska Symposium on Motivation (D. Levine, editor), Lincoln, Univ. of Nebraska Press, 1964, pp. 221–273.
8 SCHACHTER, S. AND J. E. SINGER. Cognitive, social, and physiological determinants of emotional state, *Psychol. Rev.*, 1962, Vol. 69, pp. 379–399.
9 SIDMAN, M., J. W. MASON, J. V. BRADY, AND J. THACH, JR. Quantitative rela-

tions between avoidance behavior and pituitary-adrenal cortical activity, *J. Exp. Anal. Behav.*, 1962, Vol. 5, pp. 353–362.

SEYMOUR S. KETY. *Psychoendocrine Systems and Emotion: Biological Aspects*

1. AXELROD, J. Metabolism of epinephrine and other sympathomimetic amines, *Physiol. Rev.*, 1959, Vol. 39, pp. 751–776.
2. BLOOM, G., U. S. VON EULER, AND M. FRANKENHAEUSER. Catecholamine excretion and personality traits in paratroop trainees, *Acta Physiol. Scand.*, 1963, Vol. 58, pp. 77–89.
3. ELMADJIAN, F., J. M. HOPE, AND E. T. LAMSON. Excretion of epinephrine and norepinephrine in various emotional states, *J. Clin. Endocrinol. and Metab.*, 1957, Vol. 17, pp. 608–620.
4. FOLKOW, B. AND U. S. VON EULER. Selective activation of noradrenaline and adrenaline producing cells in the cat's adrenal gland by hypothalamic stimulation, *Circulation Res.*, 1954, Vol. 2, pp. 191–195.
5. KETY, S. S. Amino acids, amines and behavior, *in* Ultrastructure and Metabolism of the Nervous System (S. R. Korey, A. Pope and E. Robbins, editors), Vol. XL, *Res. Publ., Assoc. Res. Nervous Mental Disease*, Baltimore, Williams and Wilkins, 1962, pp. 311–324.
6. KETY, S. S. Biochemistry and mental function, *Nature*, 1965, Vol. 208, pp. 1252–1257.
7. KOPIN, I. J. AND E. K. GORDON. Metabolism of norepinephrine-H^3 released by tyramine and reserpine, *J. Pharmacol. and Exp. Therap.*, 1962, Vol. 138, pp. 351–359.
8. LABROSSE, E. H., J. AXELROD, I. J. KOPIN, AND S. S. KETY. Metabolism of 7-H^3-epinephrine-d-bitartrate in normal young men, *J. Clin. Invest.*, 1961, No. 1, pp. 253–260.
9. LEVI, L. The stress of everyday work as reflected in productiveness, subjective feelings and urinary output of adrenaline and noradrenaline under salaried and piece-work conditions, *J. Psychosomat. Res.*, 1964, Vol. 8, pp. 199–202.
10. LEVI, L. The urinary output of adrenaline and noradrenaline during pleasant and unpleasant emotional states: A preliminary report, *Psychosomat. Med.*, 1965, Vol. 27, No. 1, pp. 80–85.
11. SCHACHTER, S. AND J. E. SINGER. Cognitive, social, and physiological determinants of emotional state, *Psychol. Rev.*, 1962, Vol. 69, pp. 379–399.
12. TOLSON, W. W., J. W. MASON, E. J. SACHAR, D. A. HAMBURG, J. H. HANDLON, AND J. R. FISHMAN. Urinary catecholamine responses associated with hospital admission in normal human subjects, *J. Psychosomat. Res.*, 1965, Vol. 8, pp. 365–372.
13. WURTMAN, R. J. AND J. AXELROD. Adrenaline synthesis: control by the pituitary gland and adrenal glucocorticoids, *Science*, 1965, Vol. 150, pp. 1464–1465.

D. W. WOOLLEY. *Involvement of the Hormone Serotonin in Emotion and Mind*

1 PARE, C. M. B., M. SANDLER, AND R. S. STACEY. Hydroxytryptamine deficiency in phenylketonuria, *Lancet,* 1957, Vol. 1, pp. 551–553.
2 SHORE, P. A., A. PLETSCHER, E. G. TOMICH, A. CARLSON, R. KUNTZMAN, AND B. B. BRODIE. Role of brain serotonin in reserpine action, *Ann. N.Y. Acad. Sci.,* 1956, Vol. 66, pp. 609–615.
3 Symposium on amine oxidase inhibitors, *Ann. N.Y. Acad. Sci.,* 1959, Vol. 80, pp. 553–1046.
4 WOOLLEY, D. W. Biochemical Bases of Psychoses, New York, Wiley, 1962.
5 WOOLLEY, D. W. AND B. W. GOMMI. Serotonin receptors, VII. Activities of various pure gangliosides as the receptors, *Proc. Natl. Acad. Sci.,* 1965, Vol. 53, pp. 959–963.
6 WOOLLEY, D. W. AND E. N. SHAW. A biochemical and pharmacological suggestion about certain mental disorders, *Proc. Natl. Acad. Sci.,* 1954, Vol. 40, pp. 228–231.
7 WOOLLEY, D. W. AND E. N. SHAW. Some neurophysiological aspects of serotonin, *Brit. Med. J.,* 1954, Vol. 2, pp. 122–126.
8 WOOLLEY, D. W. AND T. VAN DER HOEVEN. Prevention of a mental defect of phenylketonuria with serotonin congeners such as melatonin or hydroxytryptophan, *Science,* 1964, Vol. 144, pp. 1593–94.

STANLEY SCHACHTER. *Cognitive Effects on Bodily Functioning Studies of Obesity and Eating*

1 BEEBE, L. The Big Spenders, New York, Doubleday, 1966.
2 BLISS, E. L. AND C. H. BRANCH. Anorexia Nervosa, New York, Paul B. Hoeber, 1960.
3 BRUCH, HILDE. Transformation of oral impulses in eating disorders: a conceptual approach, *Psychiat. Quart.,* 1961, Vol. 35, No. 3, pp. 458–481.
4 CANNON, W. B. Bodily Changes in Pain, Hunger, Fear and Rage, New York, D. Appleton, 1915.
5 CARLSON, A. J. The Control of Hunger in Health and Disease, Chicago, Univ. of Chicago Press, 1916.
6 GOLDMAN, R., M. JAFFA, AND S. SCHACHTER. Yom Kipper, Air France, dormitory food and the eating behavior of obese and normal persons. Unpublished manuscript, 1967.
7 HASHIM, S. A. AND T. B. VAN ITALLIE. Studies in normal and obese subjects with a monitored food dispensing device, *Ann. N.Y. Acad. Sci.,* 1965, Vol. 131, Art. 1, pp. 654–661.
8 MARAÑON, G. Contribution à l'étude de l'action émotive de l'adrénaline, *Revue Française D'Endocrinologia,* 1924, Vol. 2, No. 5, pp. 301–325.

9 NISBETT, R. E. AND S. SCHACHTER. The cognitive manipulation of pain, *J. Exp. Soc. Psychol.*, 1966, Vol. 2, No. 3, pp. 227–236.
10 NISBETT, R. E. Taste, deprivation and weight determinants of eating behavior. Unpublished doctoral dissertation, Columbia University, 1966.
11 RAZRAN, G. The observable unconscious and the inferable conscious in current Soviet psychophysiology, *Psychol. Rev.*, 1961, Vol. 68, No. 2, pp. 81–147.
12 SCHACHTER, S. The Psychology of Affiliation, Stanford, Stanford Univ. Press, 1959.
13 SCHACHTER, S. The interaction of cognitive and physiological determinants of emotional state, *in* Advances in Experimental Social Psychology, Vol. I (L. Berkowitz, editor), New York, Academic Press, 1964, pp. 49–80.
14 SCHACHTER, S., R. GOLDMAN, AND A. GORDON. The effects of fear, food deprivation and obesity on eating. Unpublished manuscript, 1967.
15 SCHACHTER, S. AND L. GROSS. Manipulated time and eating behavior. Unpublished manuscript, 1967.
16 SCHACHTER, S. AND B. LATANE. Crime, cognition and the autonomic nervous system, *in* Nebraska Symposium on Motivation (D. Levine, editor), Lincoln, Univ. of Nebraska Press, 1964, pp. 221–273.
17 SCHACHTER, S. AND J. SINGER. Cognitive, social, and physiological determinants of emotional state, *Psychol. Rev.*, 1962, Vol. 69, No. 5, pp. 379–399.
18 SCHACHTER, S. AND L. WHEELER. Epinephrine, chlorpromazine, and amusement, *J. Abn. and Soc. Psychol.*, 1962, Vol. 65, No. 2, pp. 121–128.
19 STUNKARD, A. Obesity and the denial of hunger, *Psychosomat. Med.*, 1959, Vol. 21, No. 4, pp. 281–289.
20 STUNKARD, A. Hunger and satiety, *Am. J. Psychiat.*, 1961, Vol. 118, No. 3, pp. 212–217.
21 STUNKARD, A. AND C. KOCH. The interpretation of gastric motility: I. Apparent bias in the reports of hunger by obese persons, *Arch. Gen. Psychiat.*, 1964, Vol. 11, pp. 74–82.

ACKNOWLEDGMENT

Much of the research described in this paper was supported by Grant MH 05203 from the National Institute of Mental Health, United States Public Health Service, and by Grants G23758 and GS 732 from the National Science Foundation.

MARVIN STEIN. *Some Psychophysiological Considerations of the Relationship Between the Autonomic Nervous System and Behavior*

1 AHLQUIST, R. P. A study of the adrenotropic receptors, *Am. J. Physiol.*, 1948, Vol. 153, pp. 586–600.
2 BROBECK, J. R. Regulation of feeding and drinking, *in* Handbook of Physiology, Section 1: Neurophysiology, Vol. 2 (H. W. Magoun, editor), Washington, D.C., American Physiological Soc., 1960, pp. 1197–1206.
3 CANNON, W. B. AND J. E. URIDIL. Studies on the conditions of activity in endo-

crine glands. VIII. Some effects on the denervated heart of stimulating the nerves of the liver, *Am. J. Physiol.,* 1921–22, Vol. 58, pp. 353–364.

4 CANNON, W. B. AND A. L. WASHBURN. An explanation of hunger, *Am. J. Physiol.,* 1912, Vol. 29, pp. 441–454.

5 CARLSON, A. J. The Control of Hunger in Health and Disease, Chicago, Univ. of Chicago Press, 1916.

6 DAVIS, R. C. Response patterns, *Trans. N.Y. Acad. Sci.,* 1957, Vol. 19, pp. 731–739.

7 DIXON, W. E. AND P. HAMILL. The mode of action of specific substances with special reference to secretin, *J. Physiol.,* 1909, Vol. 38, pp. 314–331.

8 ELLIOTT, T. R. The action of adrenalin, *J. Physiol.,* 1905, Vol. 32, pp. 401–467.

9 EULER, U. S. VON. A specific sympathomimetic ergone in adrenergic nerve fibres (sympathin) and its relations to adrenaline and nor-adrenaline. *Acta. Physiol. Scand.,* 1946, Vol. 12, pp. 73–97.

10 EULER, U. S. VON. Autonomic neuroeffector transmission, *in* Handbook of Physiology, Section 1: Neurophysiology, Vol. I, Washington, D.C., American Physiological Soc., 1959, pp. 215–237.

11 GADDUM, J. H. AND L. G. GOODWIN. Experiments on liver sympathin, *J. Physiol.,* 1947, Vol. 105, pp. 357–369.

12 HAGEN, P. The storage and release of catecholamines, *Pharmacol. Rev.,* 1959, Vol. 11, pp. 361–373.

13 HAGEN, J. H. AND P. B. HAGEN. Actions of adrenalin and noradrenalin on metabolic systems, *in* Actions of Hormones on Molecular Processes (G. Litwack and D. Kritchevsky, editors), New York, Wiley, 1964, pp. 268–319.

14 HUNT, W. A. AND H. CANTRIL. Emotional effects produced by the injection of adrenalin, *Am. J. Psychol.,* 1932, Vol. 44, No. 2, pp. 300–307.

15 KRON, R. E., M. STEIN, K. GODDARD, AND M. D. PHOENIX. The effect of nutrient upon the sucking behavior of newborn infants, *Psychosomat. Med.,* 1967, Vol. 29, No. 1, p. 24.

16 LACEY, J. I. Psychophysiological approaches to the evaluation of psychotherapeutic process and outcome, *in* Research in Psychotherapy (Eli A. Rubinstein and Morris B. Parloff, editors), Washington, D.C., Am. Psychol. Assoc., 1959, pp. 160–208.

17 LACEY, J. I. AND B. C. LACEY. Verification and extension of the principle of autonomic response-sterotypy, *Am. J. Psychol.,* 1958, Vol. 71, pp. 50–73.

18 MAYER, S. E. AND N. C. MORAN. Relation between pharmacologic augmentation of cardiac contractile force and the activation of myocardial glycogen phosphorylase, *J. Pharmacol. Exp. Therap.,* 1960, Vol. 129, pp. 271–281.

19 MILLER, N. E. Experiments on motivation, *Science,* 1957, Vol. 126, pp. 1271–1278.

20 POWELL, C. E. AND I. H. SLATER. Blocking of inhibitory adrenergic receptors by a dichloro analog of isoproternol, *J. Pharmacol. Exp. Therap.,* 1958, Vol. 122, pp. 480–488.

21 ROTHBALLER, A. B. The effects of catecholamines on the central nervous system, *Pharmacol. Rev.,* 1959, Vol. 11, pp. 494–547.

22 SCHACHTER, S. AND J. SINGER. Cognitive, social, and physiological determinants of emotional state, *Psychol. Rev.*, 1962, Vol. 69, pp. 379–399.
23 SCHIAVI, R. C., M. STEIN, AND B. B. SETHI. Respiratory variables in a response to a pain-fear stimulus and in experimental asthma, *Psychosomat. Med.*, 1961, Vol. 23, No. 6, pp. 485–492.
24 SHERRINGTON, C. S. Cutaneous sensations, *in* Textbook of Physiology, Vol. 2 (E. A. Shäfer, editor), Edinburgh, Pentland, 1900, pp. 920–1001.
25 STEIN, M. Etiology and mechanisms in the etiology of asthma, *in* Psychosomatic Medicine: The First Hahnemann Symposium (John Hazen Rodine and J. H. Moyer, editors), Philadelphia, Lea and Febiger, 1962, pp. 149–156.
26 STEIN, M., R. SCHIAVI, P. OTTENBERG, AND C. HAMILTON. The mechanical properties of the lungs in experimental asthma in the guinea pig, *J. Allergy*, 1961, Vol. 32, pp. 8–16.
27 STUNKARD, A. Obesity and the denial of hunger, *Psychosomat. Med.*, 1959, Vol. 21, No. 4, pp. 281–289.
28 VANE, J. R. Catechol Amines, *in* Recent Advances in Pharmacology (3rd ed.) (J. M. Robson and R. S. Stacey, editors), Boston, Little, Brown, 1962, pp. 95–121.
29 WAN, W. AND M. STEIN. Unpublished observations.
30 WATTS, D. T. AND T. R. POOLE. Peripheral blood epinephrine levels following intravenous administration of the material, *Federation Proc.*, 1957, Vol. 16, p. 344.

NORMAN A. SCOTCH. *Inside Every Fat Man*

1 KOOS, EARL L. The Health of Regionville, New York, Columbia Univ. Press, 1954.
2 RAY, VERNE F. Techniques and problems in the study of human color perception, S. W. J. Anthrop. *J. Anthrop.*, 1952, Vol. 8, No. 3, pp. 251–259.
3 RAY, VERNE F. Human color perception and behavioral response, *Trans. N.Y. Acad. of Sci.*, 1953, Vol. 16, No. 2, pp. 98–104.
4 SEGALL, M. H., D. T. CAMPBELL, AND M. J. HERSKOVITS. The Influence of Culture on Visual Perception, New York, Bobbs-Merrill, 1966.
5 ZBOROWSKI, MARK. Cultural components in response to pain, *J. Social Issues*, 1952, Vol. 8, No. 4, pp. 16–30.
6 ZOLA, IRVING K. Culture and Symptoms — An Analysis of Patients Presenting Complaints, *Am. Sociol. Rev.*, Vol. 31, No. 5, Oct. 1966, pp. 615–630.

V. H. DENENBERG. *Stimulation in Infancy, Emotional Reactivity, and Exploratory Behavior*

1 ADER, R. Effects of early experience and differential housing on behavior and susceptibility to gastric erosions in the rat, *J. Comp. Physiol. Psychol.*, 1965, Vol. 60, pp. 233–238.

2 ADER, R. AND M. L. BELFER. Prenatal maternal anxiety and offspring emotionality in the rat, *Psychol. Rep.*, 1962, Vol. 10, pp. 711–718.

3 ADER, R. AND P. M. CONKLIN. Handling of pregnant rats: Effects on emotionality of their offspring, *Science*, 1963, Vol. 142, pp. 411–412.

4 ADER, R., A. KREUTNER, JR., AND H. L. JACOBS. Social environment, emotionality, and alloxan diabetes in the rat, *Psychosom. Med.*, 1963, Vol. 25, No. 1, pp. 60–68.

5 BROADHURST, P. L. Emotionality and the Yerkes-Dodson Law, *J. Exp. Psychol.*, 1957, Vol. 54, pp. 345–352.

6 BROADHURST, P. L. Analysis of maternal effects in the inheritance of behaviour, *Anim. Behav.*, 1961, Vol. 9, Nos. 3 and 4, pp. 129–141.

7 CASLER, L. The effects of extra tactile stimulation on a group of institutionalized infants, *Genet. Psychol. Monogr.*, 1965, Vol. 71, pp. 137–175.

8 DENELSKY, G. The influence of early experience upon adult exploratory behavior. Unpublished Ph.D. dissertation, Purdue University, 1966.

8a DENELSKY, G. Y. AND V. H. DENENBERG. Infantile stimulation and adult exploratory behavior: effects of handling upon tactual variation-seeking, *J. Comp. Physiol. Psychol.*, 1967, Vol. 63, pp. 309–312.

8b DENELSKY, G. Y. AND V. H. DENENBERG. Infantile stimulation and adult exploratory behaviour in the rat: effects of handling upon visual variation-seeking, *Anim. Behav.*, 1967 (in press).

9 DENENBERG, V. H. Critical periods, stimulus input, and emotional reactivity: A theory of infantile stimulation, *Psychol. Rev.*, 1964, Vol. 71, pp. 335–351.

10 DENENBERG, V. H. Animal studies on developmental determinants of behavioral adaptability, in Experience, Structure, and Adaptability (O. J. Harvey, editor), New York, Springer, 1966, pp. 123–147.

11 DENENBERG, V. H., P. V. CARLSON, AND M. W. STEPHENS. Effects of infantile shock upon emotionality at weaning, *J. Comp. Physiol. Psychol.*, 1962, Vol. 55, No. 5, pp. 819–820.

12 DENENBERG, V. H. AND L. J. GROTA. Social-seeking and novelty-seeking behavior as a function of differential rearing histories, *J. Abn. Soc. Psychol.*, 1964, Vol. 69, pp. 453–456.

13 DENENBERG, V. H. AND G. G. KARAS. Effects of differential infantile handling upon weight gain and mortality in the rat and mouse, *Science*, 1959, Vol. 130, pp. 629–630.

14 DENENBERG, V. H. AND G. G. KARAS. Interactive effects of infantile and adult experiences upon weight gain and mortality in the rat, *J. Comp. Physiol. Psychol.*, 1961, Vol. 54, pp. 685–689.

15 DENENBERG, V. H. AND J. R. C. MORTON. Effects of preweaning and postweaning manipulations upon problem-solving behavior, *J. Comp. Physiol. Psychol.*, 1962, Vol. 55, No. 6, pp. 1096–1098.

16 DENENBERG, V. H. AND J. R. C. MORTON. Infantile stimulation, prepubertal sexual-social interaction and emotionality, *Anim. Behav.*, 1964, Vol. 12, pp. 11–13.

17 DENENBERG, V. H., J. R. C. MORTON, AND G. C. HALTMEYER. Effects of social

groupings upon emotional behavior, *Anim. Behav.*, 1964, Vol. 12, pp. 205–208.

18 DENENBERG, V. H., J. R. C. MORTON, N. J. KLINE, AND L. J. GROTA. Effects of duration of infantile stimulation upon emotionality, *Can. J. Psychol.*, 1962, Vol. 16, No. 1, pp. 72–76.

19 DENENBERG, V. H., D. R. OTTINGER, AND M. W. STEPHENS. Effects of maternal factors upon growth and behavior of the rat, *Child Develop.*, 1962, Vol. 33, pp. 65–71.

20 DENENBERG, V. H. AND S. A. SMITH. Effects of infantile stimulation and age upon behavior, *J. Comp. Physiol. Psychol.*, 1963, Vol. 56, pp. 307–312.

21 DENENBERG, V. H. AND A. E. WHIMBEY. Behavior of adult rats is modified by the experiences their mothers had as infants, *Science*, 1963, Vol. 142, pp. 1192–1193.

22 GAURON, E. F. AND W. C. BECKER. The effects of early sensory deprivation on adult rat behavior under competition stress: an attempt at replication of a study by Alexander Wolf, *J. Comp. Physiol. Psychol.*, 1959, Vol. 52, No. 6, pp. 689–693.

23 HEYNS, O. S. Abdominal Decompression, Johannesburg, South Africa, Witwatersrand Univ. Press, 1963.

24 HEYNS, O. S., J. M. SAMSON, AND J. A. C. GRAHAM. Influence of abdominal decompression on intra-amniotic pressure and foetal oxygenation, *Lancet*, 1962, Vol. 1, pp. 289–292.

25 HEYNS, O. S., J. M. SAMSON, AND W. A. B. ROBERTS. An analysis of infants whose mothers had decompression during pregnancy, *Med. Proc.*, 1962, Vol. 8, Part 2, pp. 307–311.

26 HOCKMAN, C. H. Prenatal maternal stress in the rat: its effects on emotional behavior in the offspring, *J. Comp. Physiol. Psychol.*, 1961, Vol. 54, No. 6, pp. 679–684.

27 HUNT, H. F. AND L. S. OTIS. Early "experience" and its effects on later behavioral processes in rats: 1. Initial experiments, *Trans. N.Y. Acad. Sci.*, 1963, Vol. 25, pp. 858–870.

28 KARAS, G. G. AND V. H. DENENBERG. The effects of duration and distribution of infantile experience on adult learning, *J. Comp. Physiol. Psychol.*, 1961, Vol. 54, pp. 170–174.

29 LANDAUER, T. K. AND J. W. M. WHITING. Infantile stimulation and adult stature of human males, *Amer. Anthropologist*, 1964, Vol. 66, pp. 1007–1028.

30 LEVINE, S. A further study of infantile handling and adult avoidance learning, *J. Pers.*, 1956, Vol. 25, No. 1, pp. 70–80.

31 LEVINE, S. Infantile experience and consummatory behavior in adulthood, *J. Comp. Physiol. Psychol.*, 1957, Vol. 50, pp. 609–612.

32 LEVINE, S. Noxious stimulation in infant and adult rats and consummatory behavior, *J. Comp. Physiol. Psychol.*, 1958, Vol. 51, pp. 230–233.

33 LEVINE, S. The psychophysiological effects of infantile stimulation, *in* Roots of Behavior (Eugene L. Bliss, editor), New York, Harper, 1962, pp. 246–253.

34 LINDHOLM, B. W. Critical periods and the effects of early shock on later emo-

tional behavior in the white rat, *J. Comp. Physiol. Psychol.*, 1962, Vol. 55, No. 4, pp. 597–599.
35 MORRA, M. Level of maternal stress during two pregnancy periods on rat offspring behavior, *Psychonom. Sci.,* 1965, Vol. 3, pp. 7–8.
36 MORTON, J. R. C. The interactive effects of preweaning and postweaning environments upon adult behavior. Unpublished Ph.D. dissertation, Purdue University, 1962.
37 MORTON, J. R. C., V. H. DENENBERG, AND M. X. ZARROW. Modification of sexual development through stimulation in infancy, *Endocrinology,* 1963, Vol. 72, pp. 439–442.
38 OTTINGER, D. R., V. H. DENENBERG, AND M. W. STEPHENS. Maternal emotionality, multiple mothering, and emotionality in maturity, *J. Comp. Physiol. Psychol.*, 1963, Vol. 56, pp. 313–317.
39 OTTINGER, D. R. AND J. E. SIMMONS. Behavior of human neonates and prenatal maternal anxiety, *Psychol. Rept.*, 1964, Vol. 14, pp. 391–394.
40 STERN, J. A., G. WINOKUR, A. EISENSTEIN, R. TAYLOR, AND M. SLY. The effect of group vs. individual housing on behavior and physiological responses to stress in the albino rat, *J. Psychosomat. Res.*, 1960, Vol. 4, pp. 185–190.
41 THOMPSON, W. R. Influence of prenatal maternal anxiety on emotionality in young rats, *Science,* 1957, Vol. 125, pp. 698–699.
42 WAJDOWICZ, E. K. Abdominal decompression during labor, *Amer. J. Nursing,* 1964, Vol. 64, No. 12, pp. 87–89.
43 WHIMBEY, A. E. The factor structure underlying the experimentally created individual differences studied in "early experience" research. Unpublished Ph.D. dissertation, Purdue University, 1965.
44 WHIMBEY, A. E. AND V. H. DENENBERG. Experimental programming of life histories: The factor structure underlying experimentally created individual differences, *Behaviour,* 1968 (in press).
45 WHIMBEY, A. E. AND V. H. DENENBERG. Programming life histories: Creating individual differences by the experimental control of early experiences, *Multivariate Behavioral Research,* 1966, Vol. 1, pp. 279–286.
46 WHITE, B. L. AND P. W. CASTLE. Visual exploratory behavior following postnatal handling of human infants, *Percept. Mot. Skills,* 1964, Vol. 18, pp. 497–502.
47 WHITING, J. W. M. Menarcheal age and infant stress in humans, *in* Sex and Behavior (F. A. Beach, editor), New York, Wiley, 1965, pp. 221–233.
48 WOLF, A. The dynamics of the selective inhibition of specific functions in neurosis: A preliminary report, *Psychosomat. Med.*, 1943, Vol. 5, No. 1, pp. 27–38.

ACKNOWLEDGMENT

Much of the author's research described in this paper has been supported by grants from the National Institutes of Health and the Purdue Research Foundation.

J. P. Scott. *Biology and the Emotions*

1. Bolles, R. C. and P. J. Woods. The ontogeny of behaviour in the albino rat, *Animal Behav.*, 1964, Vol. 12, No. 4, pp. 427–441.
2. Denenberg, V. H. The effects of early experience, *in* The Behaviour of Domestic Animals (E. S. E. Hafez, editor), Baltimore, Williams and Wilkins, 1962, pages 109–138.
3. Hall, C. S. Emotional behavior in the rat, *J. Comp. Psychol.*, 1934, Vol. 18, No. 3, pp. 385–403.
4. Levine, S. Psychophysiological effects of infantile stimulation, *in* Roots of Behavior (Eugene L. Bliss, editor), New York, Harper, 1962, pp. 246–253.
5. Pasamanick, B. and H. Knobloch. Epidemiologic studies on the complications of pregnancy and the birth process, *in* Prevention of Mental Disorder in Children (G. Caplan, editor), New York, Basic Books, 1961, pages 74–94.
6. Scott, J. P. and J. L. Fuller. Genetics and the Social Behavior of the Dog, Chicago, Univ. of Chicago Press, 1965.

John W. M. Whiting. *Analysis of Infant Stimulation*

1. Denenberg, V. H. and J. R. C. Morton. Effects of preweaning and postweaning manipulations upon problem-solving behavior, *J. Comp. Physiol. Psychol.*, 1962, Vol. 55, No. 6, pp. 1096–1098.
2. Denenberg, V. H., D. R. Ottinger, and M. W. Stephens. Effects of maternal factors upon growth and behavior of the rat, *Child Develop.*, 1962, Vol. 33, pp. 65–71.
3. Denenberg, V. H. and A. E. Whimbey. Behavior of adult rats is modified by the experiences their mothers had as infants, *Science*, 1963, Vol. 142, pp. 1192–1193.
4. Gunders, S. The effects of periodic separation from the mother during infancy upon growth and development. Ed.D. thesis, 1961, Harvard University.
5. Gunders, S. and J. W. M. Whiting. The effects of periodic separation from the mother during infancy upon growth and development. Presented August, 1964, *at* International Congress of Anthropological and Ethnological Sciences, Moscow.
6. Landauer, T. K. and J. W. M. Whiting. Infantile stimulation and adult stature of human males, *Amer. Anthropologist*, 1964, Vol. 66, No. 5, pp. 1007–1028.
7. Ottinger, D. R., V. H. Denenberg, and M. W. Stephens. Maternal emotionality, multiple mothering, and emotionality in maturity, *J. Comp. Physiol. Psychol.*, 1963, Vol. 56, No. 2, pp. 313–317.

Index

Note: Numbers in italics refer to pages on which figures appear

A

ACTH
  conditioned anxiety and *80*
  plasma 17-hydroxycorticosteroids elevation and 79, *80*
activation, as indicator of neural configurational change 11
activation theories 10 ff.
adaptation
  catecholamine levels and, value of 105
  corticosteroids, value of in 106
adaptive function, concept of 190, *192*, 193
adrenergic receptors, types of 146
adrenocorticotrophic hormone (see ACTH)
affective behavior 72
amacrine cells, inhibitory processes and 16
amygdaloid complex
  effects of ablation on 23, *26*, *27*
  efferent control of inhibition and *20*, *21*, *22*, *24*, *25*
  electrical stimulation of, and attack response in cats 55, *56*, *57*, *58*
  habituation and 23, 27, *28*, *29*
  orienting reaction and 23, 27, *28*, *29*
  as reinforcer-register mechanism 28, 29, 30, *31*, *32*, *33*, *34*
anger, manipulation of and cognitive factors 122
anticipation and control, as response availability 96 ff.
anxiety, and availability of relevant behavior 100
  conditioned 76, *78*, *79*, 80, 82, 84, 85
    ACTH and *80*
    autonomic activity and 84, *85*
    catecholamines and 80, *82*
    17-hydroxycorticosteroids and 76, *78*, *79*, *80*, 81, *82*
  prenatal influence of in rats 165 ff.
  prenatal stimulation and 187
appetite, effects of hypothalamic lesions on 14
  effects of hypothalamic stimulation on 14
  finickiness and 14
  motivation and 14
arousal
  affective components and 65
  catecholamines and 146, 150
  gross 99
  manipulation of cognitive and physiological factors and 123 ff.
  physiological indexes of
    airway resistance 148, 150
    gastric motility 151
    heart rate 149
    hypoglycemia 151
    palmar sweating 149
    pulse rate 147
arousal of evaluative needs 120
asthma 148, 150
attack
  eating and 46
  effect on of cutting trigeminal nerve in cats 49, *50*, *51*
  effect on of hypothalamic stimulation 41, *45* ff.
  hypothalamic stimulation, effective sights and 54, 55
  limbic stimulation and 54, 55, *56*
  midbrain stimulation and 54, *58*, 59
  neural mechanisms of 53, 55, 59
  with rage 44, *45*, 46
  without rage 44, *45*, 46
  selectivity in
    olfaction 44, *45*, 46
    taction 51, *52*, 53

attack (cont'd)
  vision 47, 48, 51
  thalamic stimulation and 54, 56, 57, 58, 59
automated discrimination apparatus 29, 30, 31
autonomic activity
  chemical nature of 145
  differential responses and 147 ff.
  emotional conditioning and 82, 83, 84, 85
  emotions and 154
  feelings and 99
  habituation and 12
  quantitative interpretation of catecholamine effect on 104
  visceral and endocrinological regulation by 8
autonomous mechanisms 8
avoidance, conditioned 88, 89, 90, 91, 92, 93, 94
  catecholamine levels and 88, 89, 90, 92, 93, 94
  pepsinogen and 93
  17-hydroxycorticosteroids and 89, 90, 91, 92, 93

B

behavior and neurophysiology 4
biological concepts
  adaptive functions 190, 192, 193
  systems 192, 193
biological theories of emotion 4
blood levels, quantitative interpretation of 103 ff., see also ACTH, catecholamines, 17-hydroxycorticosteroids
blood pressure, and emotional conditioning 82, 83, 84, 85
blood-withdrawal procedures, effect of in emotional conditioning experiments 87, 88
brain ablations, amygdaloid complex, effects on 23, 26, 27, 28, 29, 30, 31, 32, 33, 34
  in hippocampal formation 30, 32, 33, 34
  limbic forebrain, effects on cognitive behavior 10
  sham behavior and 7

Bruch theory of obesity 126

C

Cannon-Bard theory 7, 118
care-seeking, effects of social factors on 159
catecholamine levels
  ambiguous blood-withdrawal procedures and 86, 87, 88
  autonomic activity and 104
  conditioned anxiety and 80, 82
  conditioned avoidance and 88, 89, 90, 92, 93, 94
  conditioned punishment and 86, 87
  differential release and 105
  effect on emotional states 107
  emotional conditioning and 80, 81, 82, 86, 87, 88, 89, 90, 92, 93, 94
  interaction with corticosteroid levels 106
  phenylketonuria and 114
  quantitative interpretation of 104
  response availability and 100
  tranquilizers and 113
  urine analysis and 104
catecholamines
  as adrenergic nerve transmitters 146
  antagonists of 146
  arousal and 150
  differential autonomic responses and 150
  effects on nonspecific neural systems 14
  epinephrine 145 ff.
  norepinephrine 145 ff.
cats
  amygdaloid complex in, electrical stimulation of 55, 56, 57, 58
  arousal effects of catecholamines in 147
  in brain ablation experiments, sham behavior of 7
  hippocampal formation in, electrical stimulation of 55, 56, 57, 58
  in hypothalamic stimulation experiments 41, 45 ff.
chlorpromazine, as serotonin analog 112
cognitive behavior
  electrical stimulation of limbic

cognitive behavior (cont'd)
    forebrain 10
    limbic ablation, effects on 10
cognitive effects
    on anger, experimentally manipulated 122
    differential autonomic responses and 151
    on eating behavior 121 ff., *131, 132, 134, 138, 139*
    on euphoria, experimentally manipulated 122
    on fear, experimentally manipulated 128, *132*
    on hunger, experimentally manipulated 128, *131*
    on micturition, experimentally manipulated 125
    natural, on hunger 126
    on obesity 126, *131, 132, 134,* 135
        relative to degree of overweight *139*
    on pain, experimentally manipulated 124
    social factors and 157
    on tolerance to electric shock 125
cognitive factors 107
cold pressor test 149
collateral inhibition 15, *16*, 17
    efferent control of *16, 17, 18, 19, 20, 21, 22, 24, 25*
    orienting reaction and 22 ff.
Columbia University males, in experiment on cognitive effects on eating behavior 128, *131, 132,* 142
communication, as function of emotion 191
conditioned avoidance, *see* avoidance, conditioned
conditioned emotional behavior 71 ff.
    ambiguous blood-withdrawal procedures and 87, *88*
    anxiety and 76, *78, 79,* 80, 82, 84, 85
    autonomic activity, effect on 82, *83,* 84, *85*
    avoidance and 88, *89, 90, 91, 92, 93, 94*
    hormone pattern, effect on 74, *78,*

conditioned emotional behavior (cont'd)
    79, *80, 81, 82,* 86, *87, 88, 89,* 90, *92, 93, 94*
    inflection ratio as measure of 77, *78*
    punishment and 84, *88*
    response availability and 96 ff.
    using primate restraining chair 77
conditioned punishment, *see* punishment, conditioned
configuration, in neural activity, 11
conflict, conditioned, *see* punishment, conditioned
construct of emotionality, incomplete development of 194
control of emotion 100
cortical control, of afferent inhibitory processes *16,* 17
cortiscosteroids
    differential release of 106
    interaction with catecholamines 106
critical intensity 64, *65*
critical period 197
cultural factors
    influence on body state 157
    influence on care-seeking 159
    influence on perception 158
    relation to pain 158
cybernetics and neurophysiology 4
cybernetics of emotion 35

D

differential autonomic responses, and hunger 152
distress, *see* anxiety
dogs
    development of social behavior in 196
    in experiments with serotonin 110
    and physical growth 199
dopamine 146

E

e-motion 5, 15
early experience, *see* infantile stimulation
eating behavior
    affected by taste 138, *139,* 140
    affected by time-zone changes 143
    attack and 46

eating behavior (cont'd)
  choice of eating place and 142
  cognitive factors and 121 ff., *131,
    132, 134,* 135, *138*
  differential autonomic responses
    and 152
  effect of fear on 131, *132*
  effect of formula emulsion diet on
    134
  effect of hypothalamic lesions on 14
  effect of hypothalamic stimulation
    on 15
  fasting and 141
  finickiness and 14
  motivation and 14
  obesity and 128, *131, 132, 134,* 135,
    138, 139
  time and 137, *138,* 143
  relative to degree of being over-
    weight *139*
effective behaviors 72
efferent control of input 6, *16,* 17
  by amygdaloid complex *20, 21,* 22,
    24, 25
  collateral inhibition and 17, *18, 19,
    20, 21,* 22, *24, 25*
  habituation and, neurobehavioral
    data on 23 ff.
  by inferotemporal cortex *18, 19,* 20,
    24, 25
  orienting reaction and, neurobe-
    havioral data on 23 ff.
electric shock
  to brain, and attack-response in
    cats 41, *45* ff.
    in tests of efferent control of in-
      hibition 17, *18, 19,* 20, *24, 25*
  to hypothalamus 14
  infantile stimulation in rats and
    167
  to limbic forebrain, effects on cog-
    nitive behavior 10
  to thalamus 8
  tolerance to, in experiments on cog-
    nitive effects on bodily func-
    tioning 125, 129
electrical brain responses 13, 64, *65*
  attack-response experiments with
    cats and 56
  in tests of efferent control of neural

electrical brain responses (cont'd)
  inhibition 17, *18, 19, 20, 21,* 22
electrical stimulation, *see* electric
  shock
elimination, as function of emotion
  194
emotion, derivation of word 5
emotional behavior
  adaptive functions of 190, 191, *192,
    193*
  organization of 200
emotional functions
  communication 191
  degree of reactions 191
  elimination 194
  primary stimulation 191
  reinforcement 191
  stress 191
emotional processes, neuropsychological models of 15, *16*
emotional reactivity 195
  effects of infantile social interac-
    tions on rats 168
  effects of physical infantile stimu-
    lation on rats 166, 172, *173,
    174,* 180
  effects of postweaning stimulation
    on rats 171, 180
  effects of prenatal anxiety in rats
    165 ff., 180
  factor-analysis of rats 174, *176,* 177,
    178, 179
  procedures for measuring in rats
    163
emotional states, effect of catecholamine levels on 107
emotionality, construct 194
epinephrine 145 ff.
  ambiguous blood-withdrawal pro-
    cedures and 87, *88*
  chemical action of 145
  conditioned anxiety and 80, *82*
  conditioned avoidance and 88, *89,
    90, 92, 93, 94*
  conditioned punishment and 86,
    87, *88*
  control and 100
  differential release of 105
  emotional conditioning and 80,
    81, 82, 86, 87, *88, 89,* 90, *92,*

epinephrine (cont'd)
    93, 94
  in experiments with humans and cognitive effects on bodily functioning 121 ff.
  physiological effects of 119
  quantitative interpretation of 104
  response availability and 100
ergotrophic system 8
error-evaluating mechanism, hippocampal formation as 29, *30, 31, 32, 33, 34*
euphoria
  manipulation of and cognitive factors 122
  serotonin and 113
Estes-Skinner conditioned suppression technique 76
expectation, in neural activation 11
exploratory behavior
  effects on of infantile stimulation in rats 180
  effects on of novelty stimulation in rats 181
  effects on of tactile stimulation in humans 186
  effects on of tactile stimulation in rats 182, *183*
  effects on of visual stimulation in humans 185
  effects on of visual stimulation in rats 183, *184, 185*
  factor-analysis of rats and *176*, 177, *178*, 179
expression, emotional
  as dissociated from emotion 9
  in thalamic theory 7
external control 35
external inhibition 17

F

fasting, and obesity 141
fear
  effect on hunger 131, *132*
  irrational 64
  manipulation of and cognitive factors 128, *132*
feedback 4, *5, 6*
  as homeostasis 8

feelings
  as distinguished from emotional behavior 71
  as function of emotional behavior 98
  in thalamic theory 7
finickiness
  effects of hypothalamic lesions on 14
  effects of hypothalamic stimulation on 14
flight or fight reaction, and epinephrine 106

G

galvanic skin reaction, as measure of orienting reaction 23, 27
gate control theory 66, *67, 68*

H

habituation
  amygdala control over 23, 27, *28*, 29
  effects on autonomic activity 12
  effects on visceral activity 12
  as function of self-inhibition 16
  of orienting reaction 11
  and self-inhibition, neurobehavioral data on 22 ff.
hallucinations, and serotonin 109
heart rate
  differential autonomic response tests and 149
  emotional conditioning and 82, *83*, 84, *85*
Heyns' abdominal decompression labor method 188
hippocampal formation
  ablation of 30, *32, 33, 34*
  electrical stimulation of, and attack response in cats 55, *56, 57, 58*
  as error-evaluating mechanism 29, *30, 31, 32, 33, 34*
  lesions in 29, *30, 31, 32, 33, 34*
homeostasis, as control through neural feedback 8
hormone pattern
  ambiguous blood-withdrawal procedures and 87, *88*
  conditioned anxiety and 76, *78, 79, 80*, 81

hormone pattern (cont'd)
   conditioned avoidance and 88, *89, 90, 91, 92, 93, 94*
   conditioned punishment and *86, 87, 88*
   emotional conditioning and 74, *78, 79, 80, 81, 82*, 86, *87, 88, 89, 90, 92, 93, 94*
hormones, differential release of 105
humans
   in cognitive effects on bodily function experiments 121ff.
   in hunger and gastric motility experiments 127
   in obesity experiments 128, *131, 132, 134*, 137, *138, 139*, 140
   in serotonin experiments 110
   in infantile stimulation experiments 185 ff.
   infantile stimulation of, effects on physical growth 203
   physical growth and 203
hunger
   differential autonomic responses 152
   effect of fear on 131, *132*
   gastric motility and 127
   labeling of and natural cognitive factors 126
   manipulation of and cognitive factors 128, *131*
hyperalgesia, and thalamic lesions 8
hypothalamus
   electrical stimulation of 14
      attack response, effective sites in cats and 54, *55*
      attack response in cats and 43, *45* ff.
      conditioning in cats and 43
      jaw opening in cats and 51, *52, 53*
      rage response in cats and 41, *45 ff.*
   lesions in 14
   thalamic theory and 7

I

infancy, length of in dogs 196
infantile stimulation (in humans)
   effects on physical growth 186
   effects on sexual maturation 187

infantile stimulation (cont'd)
   effects on tactile exploratory behavior 186
   effects on visual exploratory behavior 185
   experimental design and, basic criteria for 198
   intensity of 201
   nature of 201, *see also* infantile stimulation (in rats), postweaning stimulation, prenatal stimulation
infantile stimulation (in rats)
   effects on emotional reactivity, 166, 172, *173, 174*, 180
   effects on sexual maturation 187
   effects on successive generations 169, 180
   electric shock and 167
   emotional mothers and 201
   handling effects 166, 180, 201
   housing conditions, *see* social factors
   intensity of 201
   nature of 201
   sensory restriction and 167
   social interaction and 168, *see also* infantile stimulation (in humans), postweaning stimulation, prenatal stimulation
inferotemporal cortex, and efferent control of inhibition *18, 19,* 20, *24, 25*
inflection ratio, as measure of conditioned emotional behavior 77, *78*
inhibition, collateral, *see* collateral inhibition
inhibition, self, *see* self-inhibition
input 4, *5,* 6
   activation and 11 ff.
internal control 35
interruption, as related to anxiety 100
irrational fear 64
isopropylnorepinephrine 146

J

James-Lange theory 7, 118

L

LSD
  effects of 118
  as serotonin analog 110, *111*
labeling 120 ff.
  of hunger, effect of adrenalin on in obesity 132
lesions
  in amygdaloid complex 23, *26, 27*
  in hippocampus 29, *30, 31, 32, 33, 34*
  in hypothalamus 14
  in limbic forebrain, effects on memory 10
  in thalamic regions 8
limbic forebrain
  ablation of, effects on cognitive behavior 10
  efferent control and 23 ff.
  electrical stimulation of, and attack response in cats 55, *56, 57,* 58
  effects on cognitive behavior 10
  lesions in, effects on memory 10
  as seat of emotions 9

M

manipulation, effects on emotionality 202
mediating processes 61, *67, 68,* 69
medical care, effects of social and cognitive factors on 159
memory, effects of limbic lesions on 10
memory mechanisms 5
mice, in maze test of phenylketonuria with serotonin 114, *115*
micturition, manipulation of and cognitive factors 125
midbrain, electrical stimulation of, attack response in cats 54, *58,* 59
mislabeling, of hunger 127
monkeys
  in hippocampal lesion experiments 29, *30, 31, 32, 33, 34*
  in psychoendocrine experiments 74, *75,* 76 ff.
  in tests of efferent control of neural inhibition 17, *18, 19, 20, 21,* 22
motivation 6, 38, 64
  appetite and 14
multiple behavioral measures 147 ff.

N

neural organization 4, *5,* 6, 11
neurobehavioral concepts 4
neurobehavioral data, on efferent control 22 ff.
neurocybernetic concepts 4
neurological theories 4, 6 ff.
neuronal configuration 12
neurophysiological data 16
neuropsychological models of emotional processes, 15, *16*
nonspecific neural systems 12
  catecholamines, effects on 14
norepinephrine
  ambiguous blood-withdrawal procedures and 87, *88*
  chemical action of 145
  conditioned anxiety and 80, *82*
  conditioned avoidance and 88, *89, 90, 92, 93, 94*
  conditioned punishment and 86, *87, 88*
  control and 100
  differential release, of 105
  emotional conditioning and 80, *81, 82, 86, 87, 88, 89,* 90, *92, 93, 94*
  quantitative interpretation of 104
  response availability and 100

O

obesity, and choice of eating place 142
  cognitive factors and 128, *131, 132, 134, 135, 138*
  eating behavior and 128, *131, 132, 134,* 135, *138, 139*
    relative to degree of overweight *139*
  effects of hypothalamic lesions on in rats 14
  fasting and 141
  gastric motility experiments and 127
  mislabeling and 151
  mislabeling of hunger and 126, *131, 132*
  time-zone changes and 143
olfaction, and hypothalamic stimulation of attack response in cats 44, *45, 46*

organized behavior 100
orienting reaction
   amygdala control over 23, 27, *28, 29*
   collateral inhibition and 22 ff.
   as function of collateral inhibition 16
   habituation of 11
   measuring of, galvanic skin reaction 23, 27

P

pain, effects of social factors on 158
   gate control theory and 66, *67, 68*
   manipulation of and cognitive factors 124
palmar sweating, in differential autonomic response tests 149
Papez-MacLean theory 9
participatory processes of emotion 4, 5, 6, 35
passion 38
pepsinogen, and conditioned avoidance 93
perception, effects of social factors on 158
peripheral mechanisms 8
perturbation 35
phenylalanine, as emotion-related hormone inhibitor 114, *115*
phenylketonuria 114, *115*
   experimentally induced in mice 114, *115*
phrenology 61, *62*
physical growth
   effects of infantile stimulation on 186
   effects of prenatal stimulation on 188
   effects of separation of human infant on 203
   as index of consummatory behavior in dogs 199
physiologically functional psychological variables 97
   response availability as 102
physiologically nonfunctional psychological variables 98
pituitary-adrenal cortical activity, and emotional conditioning 74 ff.

postweaning stimulation, effects on emotional reactivity of rats 171, 180
prenatal stimulation
   effects on anxious behavior in humans 187
   effects on emotional reactivity in rats 165, 180
   effects on physical growth of humans 186
   genetic effects of in humans 199
   genetic effects of in rats 199
preparatory processes of emotion 4, 5, 6, 35
primary stimulation, as function of emotion 191
primate restraining chair, in psychoendocrine experiments *75, 76* ff.
programed life experiences, of rats *176, 178,* 181
psilocybin, as serotonin analog 110, *111*
psychoendocrine systems 74 ff.
psychologically functional physiological variables 97
psychologically nonfunctional physiological variables 97
pulse rate, as an index of arousal 147
punishment, conditioned *86, 87, 88*
   catecholamine levels and *86, 87*
   17-hydroxycorticosteroids and *86*

R

rage
   hypothalamic stimulation in cats and 42, *45* ff.
   real 41
   sham 40, 54
rats
   development of social behavior in 196
   effects of hypothalamic lesions on 14
   effects of hypothalamic stimulation on 15
   in experiments with salt and sugar solutions 64, 65
   in infantile stimulation experiments 162 ff. *173, 174, 176,*

rats (cont'd)
    178, 179, 186
  in programed life-experiences test *176,* 177, *178,* 179
  in Y-maze experiments 101
reactions, degree of as function of emotion 191
rebound effect 8
recovery function, effects of electrical stimulation on 17, *18, 19, 20, 21,* 22, *24, 25*
reinforcement, as function of emotion 191
reinforcer-register mechanism, amygdaloid complex as *28, 29, 30, 31, 32, 33, 34*
Renshaw interneurons 17
reserpine, as serotonin analog 11, *112*
response availability
  anxiety and 100
  epinephrine and 100
  norepinephrine and 100

S

St. Luke's Hospital, New York, study on obesity 133 ff., *134*
schizophrenia, and serotonin 109, 115
self-inhibition 15, *16,* 17
  efferent control of *16,* 17, *18, 19, 20, 21,* 22, *24, 25*
  habituation and, neurobehavioral data on 22 ff.
  postsynaptic 17
  presynaptic 17
sensory input 61
  gate control theory and 67, 68
separation of infants, effects on emotionality 202
serotonin 107, 108, *111, 112, 113,* 114
  analogs of 109, *110*
  biochemical pathways of *113,* 114
  chlorpromazine and *112*
  in experiments with live animals, 110
  LSD and 110, *111*
  phenylketonuria and 114, *115*
  psilocybin and 110, *111*
  psychometric analogs of 109, *110*
  relief of mental depression and 113
  reserpine and 111, *112*

serotonin (cont'd)
  schizophrenia and 109, 115
17-hydroxycorticosteroids 74, *78,* 79, *80, 82*
  ACTH and 79, *80*
  conditioned anxiety and 76, *78,* 79, *80,* 81, *82*
  conditioned avoidance and 89, *90, 91, 92, 93*
  conditioned punishment and *86*
17-OH-CS, *see* 17-hydroxycorticosteroids
sexual maturation, effects of infantile stimulation on 187
sham behavior 7
  in cats 40
sham-operated controls 29, *33, 34*
social behavior
  development of in dogs 196
  development of in rats 196
social-behavioral theories of emotion 4
social factors
  effects on exploratory behavior in rats 180
  emotional reactivity and, in infantile rats 168, 171, *176, 178*
  influence on body state 157
  influence on care-seeking 159
  influence on decision-making 158
  influence on perception 157
  manipulation of and emotional reactivity in rats 174, *176,* 177, *178*
  relation to pain 158
stability 5
  as affected by participatory processes 35
  as affected by preparatory processes 35
  and electrical neural activity 13
stress
  catecholamine levels and 104
  effects on infants 201
  function of emotion 191
  manipulation and 202
stressful infantile stimulation, effects on physical growth 186
systems of behavior, and associated emotions *192,* 193

T

taction, and hypothalamic stimulation of attack response in cats 51, *52, 53*
taste, and effect on eating behavior 138, *139*, 140
Test-Operate-Test Exit units 5
thalamic region, functions of 8
thalamic theory 7
thalamus
  ablation of 7
  electrical stimulation of 8
    attack response in cats 54, *58*
  lesions in 8
  as seat of emotions 7
  thalamic theory and 7
threshold of input 12
tickling 61
time, and effects on eating behavior 137, *138*
time-zone changes, effects on eating habits 143
TOTE units, *see* Test-Operate-Test Exit units
tranquilization, and serotonin 111, *112*
trigeminal nerve, effect of cutting on attack response in cats 49, *50*
trophotrophic process 8

U

urine, and catecholamine levels 104

V

visceral activity, and habituation 12
visceral brain theories 9
visceral theories 6
vision, and hypothalamic stimulation of attack response in cats *47, 48*
visual perception, effects of social factors on 158
visual recovery cycles 17, *18, 19, 20, 21, 22*

W

weaning
  in dogs 196
  in rats 196

Y

Y-maze experiments, with rats 101
Yemenite children, effects of infant-mother separation on physical growth 203
Yerkes-Dodson law 173

16685
11/68